Everyday Cheesemaking

Everyday Cheesemaking

How to Succeed Making Dairy and Nut Cheese at Home

EVERYDAY CHEESEMAKING

How to Succeed Making Dairy and Nut Cheese at Home

K. Ruby Blume

First Printing, July 1, 2014

Published by:
Microcosm Publishing
2752 N Williams Ave
Portland, OR 97227

In the DIY series

For a catalog, write or visit
MicrocosmPublishing.com

ISBN 978-1-62106-592-0

This is Microcosm #76150

Distributed in the United States and Canada by Independent Publishers Group and in Europe by Turnaround.

Edited by Lauren Hage
Designed by Joe Biel
Fonts by Ian Lynam
Cover by Meggyn Pomerleau

This book was printed on post-consumer paper by union workers in the United States.

CONTENTS

INTRODUCTION

ETHICAL CHEESE

DAIRY

Cheesemaking 101

My First Cheesemaking

The Story of Hard Cheese or Why Hard Cheese Is Hard

Everyday Cheeses From Around the World

Deep Culture

Mold Ripened Cheeses

Other Cultured Dairy Projects

NUTS & SEEDS

Everyday Non-Dairy

Milks, Creams & Sauces

Non-Dairy Yogurt

Quick & Easy Cheese

Cultured Nut Cheeses

Meltable Vegan Cheese

ENDNOTE

∽ INTRODUCTION ∾

Everyday Cheese for Everyday People

Artisan cheesemaking requires patience, technique, and either computer-controlled environments or millennia of practice handed down from generation to generation. For the rest of us holding down a job, paying the rent, and/or raising a couple kids, we probably don't have the bandwidth or the time to succeed. Many cheese projects take upwards of two days and razor sharp precision to get anything like what we expect from the store. This doesn't mean we can't make some cheese; only that we need to have realistic expectations about what is possible in a home environment for a normal person with a busy life. Everyday cheeses are those that anyone can succeed at while holding a baby in one arm, stirring the milk with the other, organizing a party, and texting with their feet. These cheeses do not require absolute sterility, exactitude of timing or temperature, or investing in fancy equipment. They are made quickly and easily in the times of milk abundance to preserve valuable proteins for leaner times. Everyday cheeses are reliable and resemble something good you'd like to put in your mouth, even when your production process is somewhat loose and free.

Why This Book?

When I decided to make cheese I was already well into a life of do-it-yourself homesteading and crafty know-how. I had been growing my own food for twenty years, and canning, brewing, and fermenting for ten. I figured cheesemaking would be pretty much like any other heirloom skill and it would be no problem to get up to speed. I purchased a home cheesemaking book, tried to follow the recipes, and failed, again and again, to create a product equal in quality and taste to what I could buy in the store. Isn't homemade supposed to be better? Many of the key techniques were not well-described in the book; nor did it contain encouraging advice for beginners, pitfalls to watch out for, or relevant troubleshooting tips.

Somewhat discouraged, I continued my research, stumbling upon a funky little website that offered a step-by-step home cheesemaking course with lots of

DIY workarounds. Hosted by a retired biology professor, Fankhauser's Cheese Page states the obvious: start with the easy stuff and build your skills. I began perusing books and websites for cheese recipes that one could, as a beginner, follow and succeed with. Around the same time, I founded a small school of urban sustainability and heirloom skills and began teaching others to make cheese.

At that time, home cheesemaking had not yet caught on. But the class sold out as fast as I could offer it. At every class, I spoke with people who had had the same experience as myself. They read a book, tried to follow a recipe, and got discouraged. So the reason for this book is simple: *I want you to succeed.*

I offer my experience and research in the hopes of making home cheesemaking a doable, pleasurable, and successful enterprise on your first attempt. You will learn about cheesemaking equipment, ingredients, and processes from a practical, non-artisan perspective as well as gaining an understanding of the many variables that can affect the outcome of your finished product. I'll give you dozens of tried and true recipes, troubleshooting tips, ways to "fix" a recipe gone wrong, and the opportunity to make and eat homemade products that taste better than you can buy at the store.

Dairy and Vegan in the Same Book? Really?

Along with increasing awareness about our food and how it is sourced, comes a diversity of choices people make about their diet. We are seeing more and more "mixed" households. Vegetarians who raise free-range birds to feed their kids, vegan dads with omnivore moms, and everything else in between. As a gluten-free omnivore, who has dabbled in the Paleolithic diet, I appreciate the option of a tasty, cultured, high-protein alternative to dairy. I also love to be able share something homemade and delectable with my vegan friends while we hash out the ethics and reasons behind our choices. Whether or not we agree in the end, most agree that some kind of cheese is a good thing. Therefor my intent is to go boldly forth and offer something for everyone in the household.

How to Use This Book

If you are going to work mainly with dairy, I encourage you to start at the beginning of the dairy section and work your way through. Once you have read about equipment and ingredients, the projects are organized from easy to more difficult and the skills build upon each other. Try each recipe at least twice before moving on. While you are working on that, read about the hard cheeses process. This section details many of the cheesemaking skills and processes you will need for later recipes and it will further your understanding of the variables that come into play in dairy cheesemaking. All the recipes may be made with cow or goat milk, or a blend of the two.

If you will be working with nuts and seeds, the projects are all similarly difficult. Read about equipment and ingredients and then start with the recipes that interest you.

If you master all of the recipes in this book, you will have quite a repertoire of cheeses. You may stop there or purchase an "artisan" cheese book and try your hand at some aged cheeses. Should you do so, consider picking one recipe and making it at least ten times. It is pretty much guaranteed that it will not come out the same way twice. Through repetition and careful record-keeping, you will slowly understand the variables and how they affect your end product.

All temperatures are in Fahrenheit. T=tablespoon, t=teaspoon, c=cup

ETHICAL CHEESE

I spent the 1980s and 1990s in the street, using giant puppets, stiltwalking, and other imagery to protest nuclear energy, the war in Iraq, the destruction of the redwood forest, racism, sexism, homophobia, and world economic injustice. After fifteen years and thousands of hours in the street, I was burnt out and disillusioned. Our efforts hardly seemed to make a dent in the injustice and inequity in the world. I took a break and spent a number of years studying permaculture and botany, learned to grow food and tend bees. I also spent time thinking and talking about what I might do to change that I could more sustainably offer. The term "localism" was newly coined and urban farming was still unheard of. But there was excitement in the air and my friends were asking me to teach them what I had been learning.

Thus, in 2008 I founded the Institute of Urban Homesteading—a homegrown folk school dedicated to "localism, resilience, and urban transformation." The school offers short, information-dense classes with the busy urban citizen in mind. Topics include organic gardening, soil health, beekeeping, urban animal husbandry, food preservation, greywater reuse, natural building, and fiber arts. My vision was to offer skill-based solutions to empower people to make small, everyday changes. Armed with information and hands-on experience, students can shift the use of resources and live lighter with more economic resilience. I wanted to help city folks be more in tune with seasonal cycles, increase awareness of where our food comes from, and reclaim heirloom skills that were lost after WWII in the rush of convenience consumerism. I had a political and environmental agenda hiding behind some crafty DIY skills.

I was just starting to explore cheesemaking. I guessed others would be excited to make healthy homemade versions of this much beloved food and thus I put together the first Cheesemaking 101 class. It was an instant hit and I could not keep up with the demand. While I was glad for the enthusiasm and the income, I quietly grumbled "cheesemaking is cool, but it is not going to save the world." As I continued to research and evolve my curriculum, I discovered that, like anything else in our modern world, changing our relationship to dairy and where it comes from has the potential to have an impact. The modern dairy, like so much else at an industrial level, is driven by profit at a cost to animals, the environment, and our health.

Rick Claymore

The goal in this section is not an in-depth exposé and analysis of the milk industry, nor a full exploration of every health concern related to milk and cheese, but a brief overview of the issues that could be important to make healthy and ethical choices about the consumption of cheese and other fermented dairy products.

The Milk Industry: Confinement Dairy and Processed Milk

The best tasting, most healthful milk and cheese comes from healthy cows eating the food their bodies were made to digest and getting plenty of fresh air and exercise. Like other whole foods, the most nutrient-dense and bio-available milk is raw, fresh, and unprocessed. Milk from healthy cows, raw and often fermented has been a staple of many traditional diets for the past 10,000 years. The people who have consumed this milk along with the meat or blood from the same animals lived long and healthy lives, free of many of our modern diseases.

These days the average person believes that raw milk is unsafe. How could this be when people lived for millennia without pasteurization? And with far fewer health problems than we have today. Pasteurization was popularized not because raw milk was unsafe, but so that dairies could be sloppy in their practices. During a period of rapid urbanization at the turn of the 19th century, urban dairies were often connected to distilleries and were called "slop dairies." Cows were fed the slop left over from grain alcohol production. These dairies were characterized by filth and horrendous conditions for the cows who had no access to their preferred food, never saw the light of day, and ate and slept in their own waste. The average life span for these animals was one year. These urban dairies were guilty of a host of sins, including adding chalk or alum to make the nutrient-poor milk whiter. Bacterial counts were high, refrigeration was not yet available to the average family and illness and death resulting from impure milk was frequent. Pasteurization was newly discovered and heavily promoted. Sickness and infant mortality went down and the dairies could continue to be as sloppy as they wished. But people didn't want pasteurization per se. "Pure milk" from cows fed on grass in the country was so coveted during this time that it cost three times more.

Up until WWII, raw dairy was upheld as the superior product and purchased by those who had access to it. People did not accept that pasteurized milk was healthier and tastier, but city people with limited financial resources bought the pasteurized slop milk because it was cheap and accessible. Industries launched food fear campaigns and lobbied hard for legislation to require pasteurization. Later, a similar phenomenon ensued around homogenization. The bottom line for both was that pasteurized and homogenized milk has a longer shelf life for shipping and selling in a grocery store.

Modern dairies are thankfully nothing like the slop dairies, but they embrace a different type of cruelty. They are the epitome of modern hygienic standards, clean, sterile automated systems where human hands hardly touch the animal, who is confined in a stall her entire life. She may be confined in a standard stall, 4' x 5'4", or a "comfort stall," which adds a whopping three to six inches in either direction. The cow is inactive, eating grain all day, hooked to a milking machine with her head in a stanchion. As a result of eating grain without exercise, the cows gain weight. This, in combination with standing on concrete all day, leads to laminitis, a condition of the hooves which causes extreme suffering. Mastitis, a lameness resulting in low milk production, is also common in these dairies. The use of antibiotics is widespread. Antibiotics pass into the milk—some 40% of our milk comes inoculated.

In these dairies, all systems are automated from feeding to milking to waste removal. The system is a closed loop where the milk goes directly from the animal through the milking machine into the processing tanks. And processed it is. Modern milk is so highly processed that it barely resembles its raw, unprocessed counterpart.

Beyond pasteurization and homogenization, the milk that comes in the jug from your supermarket, even organic milk, has been taken apart and put back together again—much as "whole wheat" flour has the wheat germ separated and then blended back in.

Once the milk leaves the animal it is separated by centrifuges into fat, protein, and various other solids and lipids. These are then recombined to standardized levels for whole, low-fat, and nonfat milks. What is left over goes on to become butter, cream, cheese, and so forth. When fat is removed, it is replaced with skimmed milk powder and enriched with chemically produced vitamins. Because it is the industry standard to do so, and because they are technically also made of "milk," dairies are not required to list these highly processed ingredients on the label. But reconstituted milk does not resemble whole milk on a molecular level and does not provide the fat necessary for our body to digest and utilize the proteins.

After the milk is put back together again, it is pasteurized at 161°, killing bacteria as well as denaturing the naturally occurring enzymes which help us digest the milk. It is stored in 6,400 gallon silos to be bottled and shipped nationwide.

The modern dairy is fully automated milk production machine.

Ethical Animal Husbandry and Integrated Agriculture

People have been domesticating and living in partnership with animals for the past 10,000 years, and regardless of your feelings on domestication, it is hard to argue with the fact that husbandry takes the needs of the animal into consideration and improves their lives, just like it is better for the health and psyche of the humans tending to them.

Traditional dairying put the grazing animals at the center of the paradigm. Pastoral people attended the animals on their seasonal migrations as they followed the trail of food. Later, as agriculture developed, animals were integrated as an important part of the system, eating agriculture waste, and offering a rich source of fertilizer and food in exchange for a level of care and protection unknown by their wild counterparts.

Industrialism brought a different set of demands to husbandry and the last hundred years have shown the ramifications of the "get big or get out" ethic: farmers enter a cycle of expenses that demand an ever-growing scale. The scale of a large dairy requires standardization, automation, and profit margins over individualized attention to animals and their needs. The result is a system where the animal is treated more like a machine than a living creature. It is doubtful that modern dairymen are indifferent to the issue, more likely that they find themselves stuck in a cycle that demands ever-greater inputs to break even, so they cut corners and care to accommodate the scale that their business requires.

In sharp contrast, the partnership model of husbandry assumes that both animal and human get something from the deal and that the best milk (meat, eggs) will come from a physically and psychologically healthy animal. This requires access to fresh air, ample space, exercise, and the right amount of the proper food for its digestive system. Small-scale traditional dairying and pasturing allows the animals a life similar to what their wild counterparts might have lived, with the benefits of more calories, healthcare, and protection. The partnership model is expensive and the people who practice it, especially small-scale farmers who provide raw dairy, are often subject to high certification costs and a measure of scrutiny not afforded their corporate counterparts.

Beyond caring for the animals as if they matter, these farmers, both traditional and modern, care about sustainable food farming systems, including soil health and longevity. They see plants and animals as integrated parts of a whole. In the current model of conventional agriculture, cows are fed cheap

GMO corn, which makes them ill, and they sit in giant feed lots where their waste off-gasses pollute the atmosphere. A few miles away, mono-cultured food crops bear the weight of many hundreds of thousands of tons of chemical fertilizers, herbicides, and pesticides. Far from enriching the soil, these petrochemicals destroy soil life and require increasing attention to function at all.

Lori Eames

Ethical livestock farmers care about the physical and psychological health of the animals and use integrative systems that restore soil fertility.

Conversely, in an integrated system, animals and plants work together in a closed loop. Wastes from plant crops are composted or feed directly to the animals who build soil fertility with their rich manure. Both compost and manure support life in the soil, which does the work of fixing nitrogen, recycling nutrients, and feeding the plants. Integrative systems work with nature, rather than against it and provide healthy nutrient dense food for both humans and animals.

Small-Scale Goat Dairy

For the last three years I have been part of a small-scale dairy collective, tending a herd of urban goats. In exchange for milking, plus weekly and monthly chores, I get a share of raw goat milk and the privilege of hanging out with these wonderful, intelligent animals. At any given time we have between three and twelve goats, usually a variety of female goats for milk and their offspring, with the occasional buck to freshen the herd.

Once a goat has been "freshened" (allowed to mate and have kids) she can give milk for up to three years. Depending on the size and breed, a single goat yields between three and eighteen gallons of milk per week. Goats are viable animals for relatively small spaces. If managed properly their poop does not need to be aged to be used in the garden and is not very smelly. For urban environments you'll want to make sure you choose an animal that is fairly quiet. Be sure to ask the seller when purchasing a goat.

Goats must be kept in pairs for companionship and need at least 10' x 15' per animal. Their living area should include shelter from wind and rain, clean bedding, and safe things to climb and play on. Goats are smart and get bored easily, so the smaller the space they are kept in, the more regularly they should be taken out for walks and allowed to meet and greet new people and environments. As well as their living area, you will need a milking area outside the goat pen, which should include a stanchion (milk stand) and a way to capture, cool, and store the milk.

A goat will clear brush if she is starving, but it is unfair to expect that from a mama goat who provides you with high-quality milk twice a day. All the goats at our small dairy get an ample supply of fresh hay and alfalfa as well as clean, fresh water. In addition, the goat receives high-quality "treats" at each milking. These include portions of organic alfalfa pellets, organic goat pellets, organic oats, kelp meal, sunflower seeds, flax meal, and probiotics. This is topped with a dollop of molasses and the occasional apple or garden treat.

In addition to living space, good food, and psychological care, raising goats requires a knowledge of health issues and access to vet resources, should anything go wrong. This in itself can be prohibitive for folks considering starting a small-scale dairy.

Beyond the benefits of milk, the small dairy offers an abundance of high-quality fertilizer, locally-sourced meat (male offspring), and companionship.

Starting and maintaining a small-scale dairy, while not impossible, is out of reach for most people. Increasingly, however, people are taking steps to reclaim food production from the agricultural-industrial complex and interest in fresh, raw, ethically-produced milk and cheese is growing. In both urban and rural places the prevalence of goat and cow shares are on the rise. People tending animals often need help, so seek out like-minded folks and ask around. A day mucking out the pen could lead to a lifetime of high-quality raw dairy!

Urban goats require regular excercise for the best physical and psychological health.

Raw Milk and the Law

Since World War II and the increasing industrialization of our food-shed, raw milk has become more and more difficult to obtain. In some states it is altogether illegal. In some it is available for pet consumption only and in others it is illegal to sell in stores but permissible through herd shares. Even if it is legal, certification usually requires industry standard equipment and processing at a cost that is prohibitive to anyone wishing to practice on a small scale.

In many cases, it is legal to consume uncertified raw milk if you own the animal and it lives on your own property; but selling, bartering, or even sharing the same milk is strictly prohibited. The herd share model is a grey-area workaround for dedicated raw milk consumers. It involves a contractual agreement with a farmer in which you purchase a "share" of the herd and pay the farmer to house and care for your animals as well as obtaining the milk for you. Crackdowns on herd shares and small dairies are on the rise, with federal agents staking out properties for months and seizing the illicit "white liquid product."

This aside, the demand for raw milk and clean food is growing and there continues to be an increase in availability, if not strictly in the legal sense, through less conventional underground channels.

Milk & Health

Milk is a high-density protein and lipid-rich food. Cheese triples or quadruples that density when you consider that one gallon of milk produces only one pound of cheese. Our bodies are made to crave fat and protein and to salivate at the sight of it. This makes good sense when you consider that for most of our history, high caloric foods were hard to come by. Most traditional dairy was consumed in fermented form, which made the milk more digestible and unlocked the nutrients through bacterial and enzymatic action.

With our current food abundance in the U.S., cheese is sold to our cravings at a rate that is far higher than what our bodies actually need. Cheese, especially cheese made from good-quality organic and/or raw milk or nuts, is an extremely delicious and nutrient abundant food. However, because of its richness, high calories and availability, you may want to consider cheese as a side portion instead of a main course.

Further, rather than consuming more of an artificially-created low-fat or non-fat dairy product, consider eating less of a high-quality whole food product. As stated earlier, reduced-fat milks are highly-processed, with natural fat and healthy cholesterol replaced with processed powdered milk proteins. The body needs fat to process protein and in absence of the necessary fats, draws on and depletes its own reserves. Oxidized cholesterol as opposed to naturally-occurring unadulterated cholesterol does initiate injury and plaque build up in the arteries.

Fresh fermented dairy products like yogurt and kefir provide the body with a fresh influx of probiotic bacteria that can aid us in digesting and utilizing milk sugars and proteins, so consider adding these to your diet several times a week along with your aged cheeses.

Milk Intolerance

Lactose, or milk sugar, is the carbohydrate portion of milk. It is broken down or digested by lactase, an enzyme produced by naturally occurring bacteria present in raw milk as well as by our own digestive system. While all baby mammals produce lactase to digest their mama's milk, humans are the only animals in which members of our species continue to produce lactase into adulthood. The people with these genes come from tribes who lived pastorally alongside dairy animals and whose genetics adapted over the last 10,000 years to be able to

utilize the valuable nutrients in milk. These people include European, Middle Eastern, East African, and Asian pastoral cultures. People who did not tend dairy animals, such as First Nation Peoples and Australian Aboriginals, stop producing lactase after age three or four. The gene for adult lactase production is dominant, so that people who mate with lactose tolerant people can pass the tolerance on to their offspring.

If you do not have the gene for adult lactase production you may have symptoms of lactose intolerance. Normally lactose would be broken down in the small intestine. If it is not, when undigested lactose reaches the large intestine, it is fermented by bacteria in the colon. This can result in intestinal cramps, bloating, and diarrhea.

Many people diagnosed as lactose intolerant can handle raw milk with no issue, since lactase is present. Fermentation also reduces the amount of lactose while boosting the amount of lactase and lactase-producing bacteria. If you are lactose intolerant you may want to test this theory, trying small amounts of raw or fermented raw milk.

Another type of milk intolerance is an intolerance or allergy to casein, one of the proteins in milk and the part of milk that goes into the cheese. The allergic reaction is caused by a casein known as Alpha S1 that is found in high levels in cow milk. The levels of Alpha S1 casein in goat milk are about 89% less, providing a far less allergenic food. Reactions to casein run from a mild intolerance, characterized by digestive issues, to a full-blown allergy including anaphylaxis. Some people may also be allergic to whey proteins, which are left in the liquid part of the milk after casein and fats have been extracted. There is no treatment for casein and whey allergies, however many people who have them, may be able to tolerate goats' milk.

Fermentation

Sandor Katz, fermentation guru and author of popular books on the topic, calls fermentation the "transformative action of micro-organisms." Bacteria and yeasts are in us, on us, and all around us. They are on the organic fruits and vegetables we eat, in the soil and in the air. The human body partners with over 1,000 micro-organisms which, for the most part, protect us and enhance our ability to live and function well. Some studies indicate that our bodies consist of more bacterial mass than human mass. Other recent studies show that people with mood disorders and autoimmune diseases are highly benefitted by increasing the amount of probiotics (beneficial bacteria) they consume.

Micro-organisms, especially lactobacilli, pre-digest a host of foods, denaturing toxins and unlocking nutrients to make them more available to our bodies. Our body can do this without the bacteria, but it taxes our system, adding stress where none is necessary. In the case of fermented dairy products, such as yogurt, kefir, quark, matsoni, and others, the proteins and fats in the milk are already partially broken down. By eating these products we also re-inoculate our system with these beneficial bacteria who can help us digest other foods. Lactobacilli (lactic acid forming bacteria) enjoy dairy, but also work on nuts, vegetables, fruits, and meats.

All traditional cultures included fermented foods in their diets and many included soured or clabbered milk. Some of these are *pima* (Finland), kefir (Slovakia, Russia), *ergo* (Ethiopia), *amasi* (South Africa), *filmjolk* (Sweden), *lassi* (Pakistan), *quark* (Germany), *matsoni* (Caspian Sea), *madila* (Botswana), and *kumis* (Mongolia).

Dairy Alternatives

While this book promotes quality dairy as a healthful food, many people have reasons to avoid dairy, which should be acknowledged and honored as intelligent, valid choices. Besides the obvious choice to avoid milk due to lactose or casein intolerance, many feel they cannot live with the injustices and problems with the milk industry or they believe it is unethical to use animal products at all.

A growing number of people have been exploring a Paleolithic diet, approximating what some of our ancestors were eating using modern equivalents. These folks believe our bodies are not evolutionarily-suited to dairy or other agricultural products, and that they are healthier without them.

Whatever your reasons for avoiding dairy products, a high-quality alternative can be found by culturing nuts and seeds. These products respond well to fermentation by lacto-bacilli and can be formed into yogurts, creams, and cheeses that are healthful and tasty. Like dairy, they offer nutrient-dense proteins and lipids. If they are cultured, raw nuts and seeds also provide beneficial enzymes.

Dairy

Cheeses

CHEESEMAKING 101

The Basics: Equipment, Ingredients, and More

The State of Cheese: Food for Thought

We live in unusual times. Never before in history has it been possible to walk into a store on any given day and choose from two hundred different styles of cheese. Historically, there were only a few cheeses available in each region. These cheeses were the result of the local temperature, humidity, and indigent bacteria. They were a direct expression of that place, expressing the *terroir* of the region (literally, "the taste of the dirt").

In order to make the two hundred or so varieties of cheese regularly available in a modern market, from Gouda to Telleme, you would need to isolate the particular strains of culture and ripening environments available naturally in their place of origin and try to recreate them in your home kitchen. In some instances this is like creating an electric greenhouse to grow mangoes in Alaska. It is not that it can't be done, but it may be more efficient, economical, and fun to simply grow regional fruits.

For the purposes of this book, we are going to work with cheeses that can easily be made in temperate North America. We'll assume that room temperature is about 72° and will try to give you options if your room temperature is higher or lower than that. Every process that is about temperature will mean paying attention to variables such as weather, season, and the ability of your pot to retain heat. Every process that is about aging is going to be affected by your local molds and bacteria. The only way you will learn what works best in your kitchen, in the summer or in the winter, is by trying it and paying attention to what works. Eventually you will hone in on the two or three cheeses that work the best and may even start to develop them to reflect your own *terroir*. At the end of this book you'll find a few recipes that urban dairymakers have developed in their own home kitchen.

Equipment You Need

To get started, there are only a few specialty items you need beyond what is already available in a well-stocked kitchen. As you go along you will discover other tools that will support you as your skills grow. A weighted, gauged cheese press is probably the most iconic piece of cheesemaking equipment, but it is actually the last thing you need to purchase, if at all.

Here's what you probably already have in your kitchen:

A pot for ripening your milk. The best pot will both conduct and hold heat well. A six, eight, or twelve-quart stainless steel pot is an excellent choice. If you have or can afford a "clad" pot, this will be best. Do not make cheese in aluminum, copper, or cast iron, which will interact with the acidifying milk.

A second pot for catching the whey.

Colander or sieve.

Measuring cups and spoons.

Things to stir with. Whisk, long stainless steel spoon, slotted spoon, rubber spatula.

Here's what you'll probably need to purchase or procure:

Measuring spoons. You probably have standard measuring spoons, but the little-bitty ones are quite useful. I have the pinch, dash, and smidgeon spoons that correspond to 1/8, 1/16, and 1/32 of a teaspoon and I use them all in cheesemaking. Find them at a high-end kitchen store or cheesemaking supply house. *(See Appendix A)*

Real cheese cloth. Not the wimpy potpourri cloth you get at the grocery, real cheese cloth has a finer weave and can be cut to size, washed, and re-used. A good grocery store might carry the cloth, as might a fabric store. Otherwise you can order from a cheesemaking supply house. *(See Appendix A)*

Curd knife. A curd knife has a long slender blade for cutting your curd. It looks a lot like a frosting spreader and can indeed be purchased in the cake-making section of your grocery or craft store.

Dairy thermometer. You can use any thermometer that has a temperature range of freezing to boiling and a clip for holding on to the side of your pot. I prefer the glass floating dairy thermometers, but some prefer a top-read dial thermometer. A frothing thermometer, such as they use to steam the milk in a café, has the correct temperature range. A candy thermometer looks similar to a glass dairy thermometer, but its range starts at 100°, so it does not cover critical temperatures for cheesemaking.

Aging container. A plastic container with a lid. I prefer one with dimensions approximately 5" x 12" x 5". They can be found at many big box or hardware stores.

Aging Container

Draining board. Some sort of latticed board that allows for drainage while keeping your cheese out of the whey. I use fluorescent light diffusers. They can be cut to any size, they lift the cheese about 3/8" above the drained whey and cost about $10 for a large sheet. Available in the lighting section of your hardware store. Cheesemaking supply houses also carry food-grade draining boards. A metal rack is not recommended as it may rust and stain the cheese.

Draining mat. Traditional draining mats are wood and look like a sushi mat. I find these get moldy and gross, so I prefer a silicon or plastic mat such as a non-slip bar mat.

Hanging hook. You will need a way to hang your cheese as it drains. A piece of wire hanging from a cabinet handle works, as does a purse hook. If you have an extra deep pot you can hang your cheese from a wooden spoon laid across the top of this pot.

Cheese molds and forms. As you learn and attempt new recipes, buy specific cheese molds or forms as you need them, rather than in anticipation of needing them.

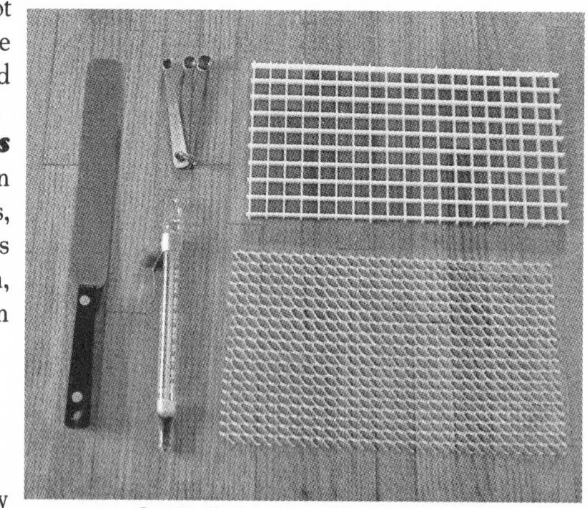

Curd knife, mini-measuring spoons, floating dairy thermometer, draining rack (above), silicon draining mat (below)

Ingredients

Milk and What is Done to It.

Source the best quality milk you can for your cheesemaking projects.

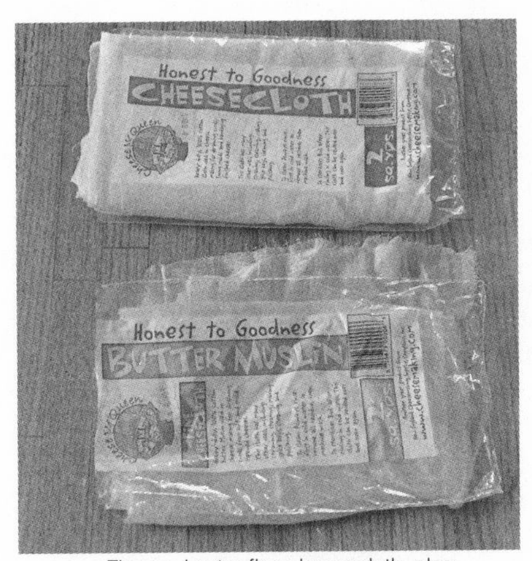

Fine and extra fine cheesecloth, also called butter muslin can be found at cheesemaking supply houses and fabric stores.

Quality milk will make the best tasting and best textured cheese. As mentioned, milk that you get in a grocery store is a highly processed item. Processing changes the molecular structure of the milk and its ability to coagulate properly. The more that has been done to make it suitable for the shelf, the more denatured it is.

Raw. The best milk for making cheese is raw milk, still warm, right from the animal. The milk comes out of the animal somewhere between 90°-95°. This is also the temperature that many cheese cultures thrive at! On the farm, it was likely that the dairy-person simply walked the fresh, warm milk to the cheese house where the local cultures literally jumped into the milk to start the cheesemaking process.

In raw milk the protein strands are unaltered which means they will coagulate well. Fresh, unprocessed milk is also full of natural enzymes. Lipase, which helps the young animal digest the fats in its mama's milk, brings a unique flavor to the cheese and can also aid human digestion of milk. Lactase, which helps the young animal access the

Food grade plastic cheese molds can be purchased online as needed. Pictured here, left to right, fresh cheese draining molds, large tomme mold with follower and small tommme mold and follower.

sugars in the milk also helps our digestive systems do the same. These flavor and digestion enhancing enzymes are denatured at about 140°.

Unless you own a goat or cow, or have access to a milk share of some sort, raw milk is precious and makes for a pretty expensive mistake if things go wrong. You may want to consider practicing on a less costly milk until you have gained some skill.

Grass Milk. Books like Michael Pollen's *Omnivores' Dilemma* and movies like *King Corn* have increased awareness of the maltreatment and malnourishment of cows, revealing that feeding grain can create suffering for the animal whose digestive system is not made to handle something so dense and rich. The grass-fed beef movement is well underway and people who care know to ask if animals are 100% grass fed or if they are "corn finished." Feeding cows their preferred diet results in meat that is leaner and tastier, not to mention the hope that the cows suffered less on their way to slaughter.

In recent months, grass milk has showed up on the shelves of my local grocery store. In a similar fashion, the milk from cows fed on grass is healthier and tastier. I can attest that Spring milk from cows pastured and grazing on green grass is absolutely divine. While the store bought version pales in comparison—it is a nice idea to think this milk might be produced with more regard for the animal's digestive health, so if cost is not a hindrance, I imagine this milk will make excellent cheese.

Cream-top milk. If you don't have access to raw milk, the next best milk for cheesemaking is cream top milk. This has been pasteurized but not homogenized. The pasteurization process heats the milk to 161° for a short time prior to bottling. This is to kill off any bacteria in the milk. Whole, organic, cream top milk is available in most natural food stores, as well as some better conventional grocery stores.

Homogenized. Homogenization is achieved by forcing the milk through an ultra-fine mesh so that the fat is broken up into particles too small to float back to the top of the milk. The homogenization process alters the structure of the milk and limits the ability for the fats and proteins to pull together well. The resulting cheese will have a rubbery texture.

Ultra-pasteurized. This milk has been heated to 240° for a sustained period of time to sterilize it such that it can sit on the shelf at room temperature without spoiling. Basically it is canned milk. Besides the fact that it tastes burned, the protein strands in ultra pasteurized milk have been weakened and

Scan the milk aisle at your supermarket or natural food store for the best quality milk you can afford.

changed so much that they will hardly pull together in the cheesemaking process.

Powdered milk plus cream. In places where cream top milk is not available, some cheese makers choose to use reconstituted powdered milk with whipping cream added, as a kind of "faux-cream top." I have heard this makes pretty good cheese.

Cow milk. High in fat and protein, good quality organic cow's milk makes excellent cheeses and is useful for a wide variety of cultured dairy projects.

Goat milk. An excellent choice for cheese making, goats' milk tends to be a bit leaner than cow milk and yields a lower amount of cheese for the same amount of milk.

Both cow and goat milk vary in richness and fat content through the year. Yields are generally lower, and milk is richer in the winter months. Conversely milk is more abundant but leaner in the summer months. Both cows and goats can stay "in milk" long after they have birthed, given that they are milked regularly once or twice per day. A mid-sized goat can give a gallon of milk a day—seven gallons a week. A standard milk cow can give up to five gallons of milk a day—up to 35 gallons of milk per week.

Sheep Milk. Delicious but rather hard to procure, as sheep do not stay in milk very long. As they are not a reliable year round source, milk is usually a sideline of having sheep for meat or fiber.

Low fat, reduced fat, 1%, 2%, and nonfat milk. As already mentioned, these milks are highly processed. Although cheese can be made from them, results will vary as a significant portion of the lipids have been replaced with powdered milk proteins. The difference in calories between whole milk and reduced fat milk is negligible and any benefit from reduction in calories is countered by a lack of fat needed to process the additional protein. So if you are counting calories you may want to consider simply eating less of a whole milk product.

The Low Down on Cheese Cultures

The bacteria used in cheesemaking (as well as sauerkraut and salami) are lactobacilli, or lactic acid forming bacteria. They seek out sugars in food and both transform and preserve the food through pre-digestion and acidification.

There are two main types of cheese cultures, mesophilic (medium temperature loving) and thermophilic (heat loving). The former work best at temperatures between 72°-100°, the latter at temperatures 90°-120°. The mesophilic cultures are mostly used in cheeses and other cultured dairy products originating from northern climates with cooler average room temperatures. The thermophilic cultures are used in yogurt and cheese originating from southern climates such as Greece and Italy with warmer ambient room temperatures.

Within these two categories, there are several species of milk-loving critters that are blended to achieve slight differences in production, mouth-feel, and flavor. There are many different series and blends of mesophilic and thermophilic cultures and this can be quite daunting for the beginner. For now, when your recipe calls for mesophilic culture, any mesophilic culture will work; when it calls for themophilic culture, any thermophilic culture will work. Freeze dried cultures for home cheesemaking are generally not available in a grocery store or even in a specialty cooking store, instead they must be purchased online from a cheesemaking supply house. If you are looking for an inexpensive DIY alternative you can obtain locally, you are in luck! When your recipe calls for mesophilic culture, you can use cultured buttermilk from the store. When your recipe calls for thermophilic culture, you can use plain yogurt from the store. If you opt for these over the counter options, use 1/4 c of buttermilk or yogurt per gallon of milk in your recipe. Once you've mastered some basics, we'll revisit cheese cultures and explore a bit deeper. See section two "Deep Culture."

Order freeze dried cultures from a cheesemaking supply house, or use over the counter alternatives from the grocer.

If you want to skip worrying about this all together, and use what you can easily obtain at the market, you are in luck! When your recipe calls for mesophilic culture, you can use cultured buttermilk from the store. When your recipe calls for thermophilic culture, you can use plain yogurt from the store. If you opt for these over the counter options, use 1/4 c of buttermilk or yogurt per gallon of milk in your recipe.

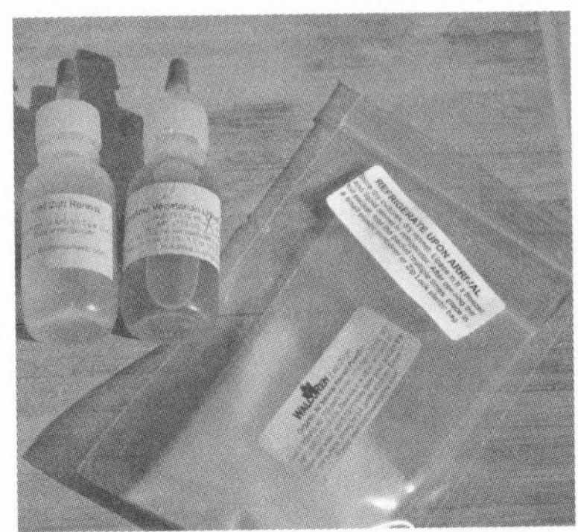

Rennet comes in an assortment of forms and strengths. Make sure you know the strength of your rennet for proper coagulation.

All Rennet is Not Created Equal

Rennet is a coagulating agent that is often used in conjunction with cultures to get a firmer curd.

Traditional rennet was made from the stomach lining of the calf and contains chymosin, an enzyme that helps the baby mammal digest its mama's milk. Chymosin helps the calf separate the proteins and lipids from the whey, exactly what we want to use it for in our cheesemaking process. Commercially purchased natural calf rennet is a by-product of the veal industry. Recently organic calf-rennet has come on the market, made by a producer in Canada.

In the U.S. today, 90% of cheese is no longer made with natural calf rennet, but with a vegetarian rennet that is GMO based. It is produced by a bacteria shot through with a cow gene. If you are vegetarian and want to use neither an animal product nor a GMO, look for organic vegetarian rennet, which we will hope and assume cannot be a GMO product.

Rennet comes in powdered, liquid, or tablet form. The liquid rennet should be kept in the refrigerator and will last about a year, at which point its strength will start to diminish, making coagulation of your projects unreliable. The powdered dry rennet is triple strength and will last indefinitely when kept in the freezer. 1/16 t of the powdered rennet is usually equivalent to 1/2 t of the liquid rennet. As some projects require an extremely dilute measure of rennet, based on drops of the weakest liquid rennet, I keep both liquid and powdered rennet on hand.

You will also find rennet in tablet forms and as junket rennet—a quarter strength rennet used in custard making. Whichever rennet you use, make sure you understand the strength and how that rennet translates to the rennet called for in your recipe. When in doubt, follow the directions on the packet of rennet, rather than in the recipe. Good quality liquid calf rennet and liquid organic vegetarian rennet are equivalent in strength.

I am often asked if there is a DIY homemade substitute for rennet. In some traditions a strong tea of nettles, cardoon, or artichoke stamens are used, however it is nigh on impossible to test the strength of this rennet for reliable results. To attempt to do so would require access to abundant free milk and a lackadaisical attitude about the high possibility of ruining a large amount of milk in the process. Should you attempt this research and have good results, please do not hesitate to report back to me!

If you want to geek out on rennet, see Appendix B.

Salt

It is not necessary to purchase a specialty "cheese salt." This is just normal salt that is processed to have fine crystals. Any non-iodized, broad spectrum salt such as sea salt or kosher salt will work for cheese making.

Coagulation: Acid Cheese vs. Rennet Cheese

Cheesemaking is the process of separating the milk solids (fat and protein) from the milk liquids (whey) through acidification. The acidification of the milk solids draws them together and also preserves them, protecting the product from opportunistic putrefying micro-organisms. The two main ways that milk is curdled to achieve this separation are:

1) a strong acid, such as lemon juice or vinegar combined with heat. This is called an acid cheese. These cheeses must be eaten within a few days of being made. They are not a living product.

2) living bacteria (culture) combined with rennet, an additional coagulating agent. These cheeses (and the whey by-product) are packed with living bacteria and some may be aged.

Basic Skills Defined

For a more detailed description of these processes see "The Story of Hard Cheese or Why Hard Cheese Is Hard" on page 51.

Figure eight stirring. Stirring in a slow figure eight motion ensures that the milk heats evenly,

Top-Stirring. Stirring with an up and down motion, like you would folding egg whites, ensures that the cultures are thoroughly mixed in.

Sanitize, Sterilize, or Just Plain Clean?

Many of the premier artisan cheesemaking books out there recommend, as your first step, that you sterilize every piece of equipment that comes into contact with your milk/cheese using a 3% chlorine bleach solution. While there is likely more benefit than harm in being this vigilant, it is not absolutely necessary to do so to succeed with the cheeses in this book and I regularly skip this step, except when attempting aged hard cheeses. In most cases a good washing with hot soapy water will do away with the bad critters we are trying to avoid.

If you should want to go one step further, you can sanitize with sodium percarbonate or other no-rinse sanitizer, available from brewing supply houses. The benefit of these sanitizing agents is that they do not have the chemical smell of bleach, nor do they require the extra step of rinsing.

One other thing to be aware of is that any wooden utensils (such as wooden spoons or spatulas with wooden handles) should be avoided or should indeed be sterilized in bleach or by boiling, as they are porous and will tend to carry bacteria—not a problem for foods and cheese you will be eating within 10 days, but a potential disaster for anything that you expect to age. Stainless steel utensils will be the most easily sanitized and the least reactive with what's in your pot.

Holding the temperature. For cheese cultures to work you will need to figure out how to hold the temperature fairly constant through the process. This can be achieved with a good quality pot, wrapping the pot in a towel to hold the heat and in extreme cases setting the pot in a warm water bath or in a warm oven (the pilot light on gas stoves or the oven light on electric stoves is probably warm enough to maintain the 90° temperature often called for).

Getting a clean break. This refers to proper coagulation of the curds. *See sidebar "Getting a Clean Break," page 44.*

Cooking the Curds. This refers to slowly bringing the temperature up on your curds and whey while stirring. The result is that the whey slowly expels from the curds, shrinking them and toughening them to prepare them to be pressed together.

Milling the cheese. Breaking up a solidified curd mass with fingers or a fork.

Draining the curds. This is achieved by lining a colander with a cheese cloth and placing a pot under it to catch the whey. The ends of the cheese cloth are then tied together and the curds are hung to drain.

YOGURT

A Curd by Any Other Namee

Yogurt is actually a very delicate curd, created by combining milk with thermophilic culture and ripening or incubating the milk at a temperature preferred by the thermophilic bacteria. It is an easy and reliable project for your first dairy culturing experiment. This recipe is scalable, up or down, and any sized jars will work, as well as recycled plastic yogurt containers.

2 large pots that can nest one inside the other
Thermometer
Mason jars or other jars with lids
Measuring cup
Spoon or whisk
Incubator *(see sidebar, page 35)*

1 gallon cow milk
1 c fresh plain cultured yogurt

1. Put the milk into the pot and re-pasteurize it by heating it on a medium heat, stirring constantly in a figure eight motion, until you reach 185°. The figure eight stirring will help to heat milk evenly and prevent burning.

2. While the milk is heating, boil water and sterilize your jars (4 quart jars plus 1 pint jar or 9 pint jars) by setting them in the boiling water (anything over 180° is fine) for 1-2 minutes each. Pull the jars out with tongs and set them upright on a clean cloth. Sterilize lids and rings as well.

3. When the milk reaches 185°, remove the milk from the heat.

4. Cool the milk to 130° in a cold water bath—set your milk into a second pot filled with cold water and stir. You can do this in a larger pot in the sink, running a stream of cold water into the bath to keep it cool. You can also add ice to your water bath. It takes less than 5 minutes to cool.

5. Combine 1 c of the cooled down milk with 1 c of plain yogurt. Sir thoroughly with a whisk to smooth out the clumps of yogurt. Add this mixture back into the pot and stir.

6. Pour into the jars and seal.

7. Incubate at 105°-115° for 3-4 hours without disturbing until set *(see sidebar, page 35)*. You can tell if it is coagulated by tipping the jar gently to the side.

Incubation Made Easy

Thermophilic cultures prefer temperatures warmer than room temperature but cooler than the lowest setting on an oven. This fact sold many, many yogurt makers in the 1970s. But clever home cheese makers have many options for incubation that do not require that purchase. Here's how:

1. Get a cheap small camping cooler. After sterilizing your jars, pour the hot water into the cooler. Before placing jars in the cooler, take a temperature reading and adjust to 122°-130°. Set your jars in the cooler. The water does not need to over the jars. Close the lid.
2. Get a cardboard box or a simple Styrofoam cooler and a 25 watt light. Cut a hole in the box and assemble the light in it. Make sure the light does not touch the box. Place your yogurt inside the box with a thermometer. Use the flaps of the box to adjust the temperature.
3. If you are lucky enough to have an old-fashioned stove with a gas pilot, your oven is probably the perfect ambient temperature. Just set your yogurt inside. Some modern stoves don't have a pilot light but the light inside the oven may warm it enough to incubate the yogurt. To test it, turn the light on inside the oven, wait an hour and read the temperature.
4. If you have an Excalibur dehydrator with a temperature gage, remove the trays, set your yogurt inside and set to 105°.
5. Some modern stoves have a dehydrator setting. Be sure to use a thermometer.

For yogurt the milk is re-pasteurized at 185°. Pictured here is a stainless steel top-read thermometer.

8. Place in the refrigerator to complete the setting process.
9. Save one pint to start your next batch. You can do this 3 or 4 times before the culture becomes contaminated or too weak to reuse—then buy fresh yogurt from the store again. Yogurt will keep 4-6 weeks in its sealed jar.

Flavoring Yogurt

We have gotten used to the convenience of pre-flavored single portion yogurts from the grocery store. In order to be sale-able with these additional ingredients, these yogurts are filled with stabilizers, gelling agents, and

sweeteners. With a few exceptions, noted below, you should flavor your yogurt when you take it out of the jar for best freshness and shelf life. Flavoring at the time of eating requires an adjustment in thinking but involves little more time and energy.

Alcohol based extracts such as vanilla or almond can be added to the milk before inoculating. Use about 2T per gallon. You can easily infuse your milk with whole vanilla bean. To do this, split the bean down the middle to expose the seeds and add it to the milk to infuse as you re-pasteurize. Strain out the bean before inoculating and filling the jars. In a similar manner, you can infuse the milk with any herbal tea or spice (think peppermint, cardamom, cinnamon). Simply add the herb or spice to the milk while re-pasteurizing and strain before inoculating.

Rob Reeves

Cool the milk in a cool water bath in the sink. Stir with a figure eight motion to bring the temperature down quickly and evenly.

All other flavoring can be done when you eat your yogurt. For honey or maple yogurt, put the yogurt in a bowl, add a dollop of honey or maple syrup and stir. For coffee yogurt just add 1/4 c coffee or espresso and a bit of sweetener. For fruit yogurts, add a bit of fresh, stewed, or preserved fruit. I enjoy canning low sugar fruit compotes all summer long so that I can eat them on my yogurt all winter.

Rob Reeves

Pouring the innoculated milk into sterile mason jars. Here half pint jars are used to be opened as a single serving. Pint and quart jars will also work, as will clean plastic containers.

Raw Milk Yogurt

In my experience it is difficult to get proper coagulation of yogurt using a raw unpasteurized milk. The resulting yogurt is usually a runny snot-like consistency. Some folks deal with this by straining the yogurt through a cloth, others use vegetarian or bovine gelatin to achieve a firmer product. I have found the former to make a yogurt that is far more tart than I like (or goaty in the case of goat yogurt) and the latter to be more like milk flavored gelatin dessert than yogurt. If you are committed to a fermented raw milk product, you may want to seek out a Kefir culture or simply let your raw milk sit at room temperature for 24-48 hours and let the naturally occurring bacteria clabber the milk.

Mesophilic Yogurt—Cultures From the Deep North

Most yogurt eaten in the U.S. is made with thermophilic cultures—there are a number of strains including those that make Bulgarian and Greek styles of yogurt. Traditionally yogurt was made at room temperature in places where room temperature was warm enough for the thermophilic critters to thrive. But there are many mesophilic yogurt-like cultures from the northern countries that work at "room temperature" (65°-75°). Some of these are *Fil Mjolk* (Scandinavia), *Matsoni* (Caspian Sea), and *Viili* (Finland). Some are quite a bit tarter or slimier than our American palates are used to, but they are worth a try as they are so simple to make and maintain. Just add a few tablespoons of the last batch to a pint of milk. Shake and leave on the counter for 24 hours. Eat. These cultures require weekly feeding and reculturing and may be found for purchase online. *(See Appendix A Resources, page 114)*.

Greek Yogurt & Yogurt Cheese

Both Greek yogurt and yogurt cheese are made by straining the yogurt through a fine cheesecloth or muslin. To make Greek-style yogurt, use a Greek yogurt as your starter culture. When your yogurt is set, line a colander with a fine cloth and pour the yogurt in. Tie the ends of the cloth and let hang 3-6 hours until you are at a consistency you enjoy. You will need to whisk or stir the yogurt to return it to a smooth consistency before eating. For yogurt cheese, continue to hang and drain for up to 12 hours. Flavor with salt and herbs, or with honey for a tasty spread. Both Greek yogurt and yogurt cheese made in this fashion will keep refrigerated for 7-10 days.

Troubleshooting Yogurt

Problem: The yogurt is a runny, snot-like consistency.

Cause: you didn't incubate long enough. The critters have not had enough time to work.

Cause: The milk is too lean. Goat milk is sometimes too lean to make good yogurt, but certain times of the year cow milk may also be too lean. Non-fat milk will also be too lean. For lean milk to get a good curd requires adding powdered milk, adding a stabilizer like gelatin or *agar agar* or straining through a doubled cheesecloth or muslin.

Cause: The temperature of your incubator was not warm enough. Check the temperature and try again.

Problem: The yogurt is grainy.

Cause: You did not temper the starter yogurt well enough. Small lumps of yogurt were stirred into the milk and curd formation was stronger around the clumps. Stir better next time.

Problem: The yogurt didn't set at all.

Cause: The yogurt culture was not viable. Get a new culture and try again.

Lori Eames

RICOTTA

Start Here Please

Ricotta is a by-product cheese first made by thrifty farm wives. Imagine an Italian farm wife somewhere deep in the Italian countryside. She has just made a wheel of Parmesan. For Parmesan to age for the two years it takes to get all that flavor and the perfect moisture content, the rind to volume ratio is very specific and the cheese needs to be large or you will end up with a dried out little rock, instead of a slightly moist flavor-packed cheese. So, in order to make that large wheel of cheese, our farmwife started with some sixty gallons of milk and a very large pot! When she drains her curds she is left with some 50 odd gallons of whey.

Being thrifty, our farm wife wants to rescue every little bit of protein from the whey before slopping it out to the pigs. So she stokes the fire under her giant pot and runs over to her vinegar crock. Of course every farm wife has a vinegar crock. This is where she throws out the little bits of old wine for the aceto-bacteria or vinegar mother to find and work their magic to turn the wine into a fine ingredient for salad dressing. So she scoops up a cup or so of the vinegar and throws it into the pot. When the pot reaches 175°, the remaining proteins in the whey pull together. She strains it out and comes up with about a quart of lovely ricotta. And she is happy, because she will make a lovely lasagna for her family.

Starting with one or two gallons of milk when we make our cheese, we could also heat the leftover whey, and add vinegar and make ricotta. We'd end up with about a tablespoon of finished ricotta, enough for a doll-sized lasagna. So instead, the following ricotta is made with whole milk, and we'll be extracting all the proteins and fats from that milk to make our cheese.

Whole Milk Ricotta

This lovely fresh tasting cheese is delicious plain, flavored or used as an ingredient in recipes. It can be used in lasagna, as a filling for blintzes, in cheesecakes, or as a spread.

2 pots
Colander
Cheese cloth
Spatula or spoon to stir

1 gallon of milk
3/4 c lemon juice

1. Put the milk into the pot and set over medium heat, stirring in a figure eight to heat milk evenly.
2. Press about 3/4 c lemon juice. Use regular lemons, not Meyer lemons.
3. When the milk reaches 175°, turn off the heat and remove from the stove.
4. Add half the lemon juice, stir, and observe. If you do not get curd formation, add half of what is remaining and stir. When the milk curdles it will do so quickly and cleanly. The whey will be a translucent yellowish color.
5. Line a colander with cheesecloth, set over a large pot and drain the curds into the cloth.
6. Tie the ends together (kitty-corner to kitty-corner) and hang the curds to drain.
7. Draining can take 10-20 minutes until it has slowed to a drip.

Ricotta, step-by-step: Curd formation

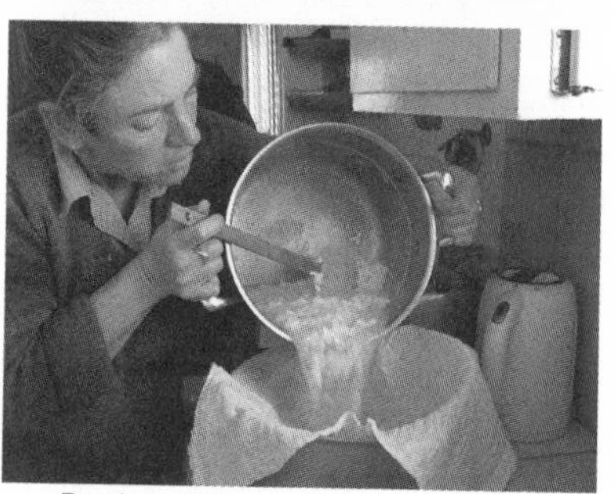

Pouring off the curds and whey into a cheesecloth lined colander

8. Open the cheesecloth and place the curds into a bowl. Break up and fluff with fingers or a fork.
9. Flavor and serve. *For flavoring suggestions see page 58.*

Hanging the curd to drain

Milling the ricotta curd mass

Variations on Ricotta

Extra creamy. Add 1/4 c of whole whipping cream to the finished cheese and mix thoroughly.

With vinegar. Substitute 1/4 c vinegar for the lemon juice. The cheese will have a slight vinegar flavor.

With citric acid. Substitute 1 t citric acid dissolved in 1/4 c water for lemon juice.

With buttermilk. Add 1 quart cultured buttermilk to the pot along with the 1 gallon of milk. Heat. The milk will curdle when it reaches 175°. This has a traditional slightly tart flavor.

Pressed. After draining, press the curds into a small bowl, ramekin, or other small container. Find a small plate that fits just inside the opening of your container and set a jar full of water onto it to press. After an hour, flip the cheese over and press again. The cheese is then sliceable.

Ricotta Salata. Similar to pressed ricotta, this cheese keeps much longer. After draining, mix the curds with 2 t salt. Press as for pressed ricotta, one hour for each side. Flip it one more time and let it sit in the mold refrigerated for 12 hours. Pull out and lightly rub the outside with salt every day for one week, returning it to the refrigerator after each rubbing. Wrap in plastic or keep in a sealed container and let it age for 2-4 weeks. This will be a very salty, crumbly cheese that can be added to a salad like feta cheese.

Troubleshooting Ricotta

Problem: The milk does not curdle.

Cause: The milk was not hot enough. Curdling will not occur until you near 175°.

Cause: The lemon was not acidic enough: You used Meyer lemons for example. Use more juice.

Cause: You did not add enough lemon juice. Add juice slowly until curdling occurs.

Problem: The cheese is too dry.

Cause: Your temperature was too high, or you added too much lemon juice. To fix you can add a little buttermilk or cream to moisten, or blend in a food processor with a little water. You could also use this batch to make pressed ricotta or ricotta salata.

Problem: The curds drain too slowly—it takes hours.

Cause: The curds are not tight enough, you did not use enough acid. You can try returning to the pot, heating and adding more acid.

FETA

Forgiving Heroine of Home Cheese Makers

Feta is the perfect everyday cheese. It is delicious on just about everything from salad to pasta and can act as a substitute for Queso Fresco in Latin dishes, for Parmesan in Italian dishes and just about everywhere else. Although not really an aged cheese, Feta keeps for quite awhile as it is stored in a salt brine. It is incredibly forgiving in its making and is thus an excellent cheese to begin with.

While many cheeses require exact temperatures during the inoculating and ripening stages, with Feta you can fudge a little here or there and still come out with a wonderful product. I have indeed made every possible mistake there is to make (wrong temperature, curds forgotten for hours, etc.), nevertheless I have always ended up with a Feta cheese. Feta is great for learning to ripen the milk, add rennet, and cut curds. There are enough variables in the process that you can learn how each affects the end product from bath to batch.

6 or 8 quart pot with lid
Thermometer
Curd Knife
Slotted spoon
Colander
Cheesecloth
Aging container
Draining rack
Muslin and elastic to cover aging container

1 gallon whole cow or goat milk
1/4 c buttermilk or 1/8 t freeze-dried mesophilic culture
1/16 t dry rennet or 1/2 tsp liquid rennet diluted in 1/4 c water
Fine iodine-free salt

1. Put the milk into the pot and heat on low to medium, stirring constantly. When you reach 90°, turn off the heat. A few degrees higher or lower will work fine and will not ruin the cheese.
2. Add mesophilic starter. Let sit for a few moments then top stir to mix the culture into the milk.

Getting a "Clean Break"

How do you know your curds are properly coagulated and you are ready to go on to the next step?

An experienced cheese maker might be able to tell by jiggling the pot or even by simply looking at the curds, but for the rest of us, using a clean finger will give us the most information. Testing for a clean break is a bit like checking to see if the cake is done. You will use one finger to pull through the curd. If it breaks cleanly over your finger—separating like a gelatin dessert and leaving a minimum of goop behind, you have a clean break.

After washing your hands, stick your index finger about 2 finger joints deep straight into the curd. Slowly rotate your wrist clockwise (or counter-clockwise if you are a lefty) to pull up through the curd. If it is ready the curd will part as if you are slicing through it. There will be a minimum of curd bits left on your finger. Additionally the break you made will be visible and a small amount of yellowish whey should pool in it. The curd will feel gelled and there will be a slight resistance (this will not be true for soft curds such as Chevre or yogurt).

If you don't think it is ready, there is no harm in letting it sit for another 5-10 minutes. If it has gone too far, the curd will pull away from

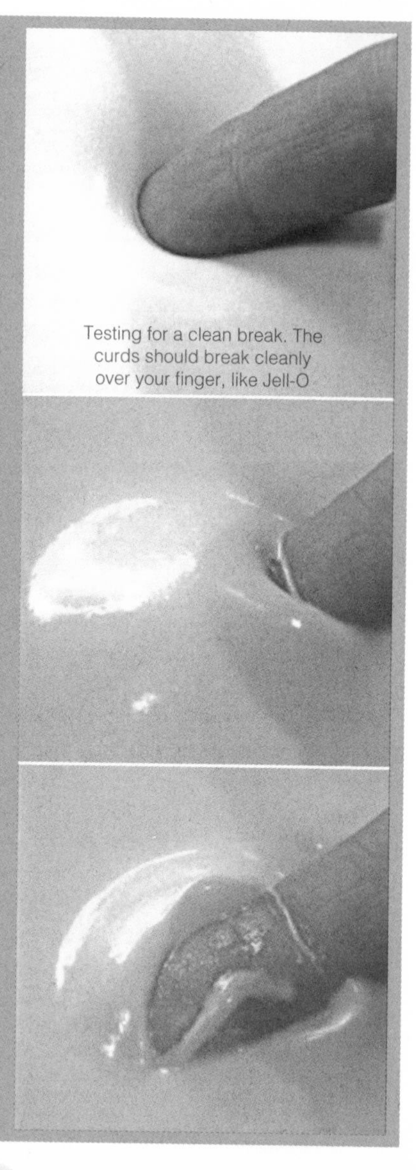

Testing for a clean break. The curds should break cleanly over your finger, like Jell-O

the side of the pot and you will have a mass of curds floating in a pool of whey. For projects in this book, you can probably still proceed with your cheese as planned. The end result will be a little firmer.

Rob Reeves

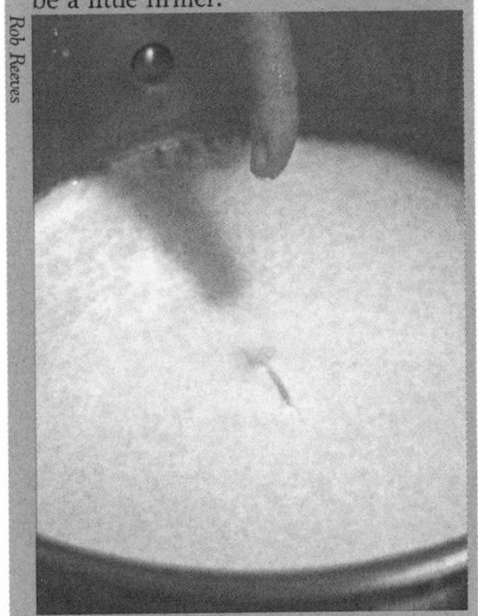

After testing for a clean break, translucent yellowish whey should pool in the break you just made in the curd.

3. Maintain the temperature at 90° for 1 hour. Depending on the quality of your pot and the ambient temperature in the room, you may want to wrap your pot in a cloth or place it in a warm spot to keep the temperature constant.

4. After one hour, add the rennet and stir 10-15 seconds.

5. Ripen 1 hour or until you get a clean break (*see sidebar*).

6. Cut the curd into 1/2 inch pieces. Send your curd knife straight down and pull cleanly across. Cut half inch slices in one direction and then in the other direction. Then set the curd knife at a 45° angle and cut through the columns of curd. Make sure you hit the bottom or the sides of the pot when cutting so you cut all the way through.

7. Maintain the 90° temperature and stir gently for 20 minutes. Use the flat of the spoon to stir so as not to cut the curds up. Use the edge of the spoon against the pot to cut any pieces that are larger than 1/2 inch. This will firm up and shrink the curds, expelling the whey. It is akin to "cooking the curds" (see page 52), although no extra heat is added.

8. Line a colander with cheesecloth and set over a pot. Drain the curds into the cheesecloth.

9. Tie the ends of the cloth, corner-to-corner and hang the curds. After 2 or 3 hours, flip the curd in the cloth (tie and re-tie) and hang another 4 to 24 hours to let the flavor develop.

10. When the cheese is ripe, take it down, unwrap it and cut the cheese into chunks. You can also cut the cheese into slabs, but I have found it harder to store that way. Lay the chunks out on the draining boards and

Rob Reeves
Rob Reeves
Rob Reeves

Making Feta, step-by-step: Cutting the curds, the curds after cutting, stirring the curds gently with the flat of the spoon, feta curds in a cheesecloth lined colander, the feta curds hanging to drain, the feta curd mass after draining, cutting the feta curd, laying the chunks of feta out on racks to be salted, feta chunks in aging container for air drying, finished feta in brine

salt lightly on all sides. You can easily flip the chunks all at once by setting a second draining rack on top of the cheeses and flipping the "sandwich."

11. Set the racks inside the aging container. Cover the container with a cloth and secure with elastic to protect from dust and flies. Let the cheese sit at room temperature 2-3 days. If room temperature is more than 80°, consider using a feta recipe that uses thermophilic culture instead. Each day drain off any whey that has collected in the bottom and flip the cheeses (again, you can do this by sandwiching them between two draining racks). The pieces will dry and develop flavor. Three or four days later, when there is less than 1/4 t whey, your cheese is ready for the next step.

12. Carefully pack cheese into a quart jar. Mix the brine by combining 1 pint water with 1 to 1.5 T salt.

Feta in Olive Oil

Feta cheese may be stored in olive oil, but should be kept cool and eaten within a month. If you add fresh herbs and /or garlic to the mixture you must refrigerate or eat within one week, as these fresh items in olive oil can promote botulism. That said this is one of my favorite ways to serve feta at a party.

1 c feta, cut into 1" chunks
1/2 c high-quality extra virgin olive oil
1/4 c sun-dried tomatoes, roasted red pepper and/or dry cured olives, sliced or diced
1 t dried basil, rosemary or other herbs of choice
1/4 t sea salt
Dash of pepper

Combine all ingredients and let marinate 24 hours. Serve with bread.

Troubleshooting Feta

Problem: Cheese is too soft or melts in the brine.
Cause: Too much moisture in the curd. Try letting your curd set longer, draining the curds longer and or adding more salt to your brine.
Problem: Cheese is too dry.
Cause: Too little moisture in the curd. Try draining your cheese for a shorter amount of time or reducing the salt in your brine.
Problem: Black or pink mold when air drying.
Cause: The room is too warm or you have let it sit too long. You can simply cut of the mold with a knife before brining.

What to Do With All That Whey

By now in your cheesemaking process you have discovered that you are left with quite a significant amount of whey when your drain your curds, and you probably feel terrible just dumping it down the drain. If you have no livestock, it is likely that at some point in your cheesemaking career you'll be dumping some whey. But should you wish to avail yourself of its usefulness, here's how:

Non-living whey from acid cheeses such as Ricotta or Panir:

1. Feed to pets or livestock. Pigs and chickens especially benefit from whey.
2. Water plants. Safe for both indoor and garden plants.
3. Add to compost.
4. Use as a base for smoothies or soup.
5. Use as the liquid in pancakes, crepes, or other batter.

Living whey from rennet cheeses, such as Feta or Chevre:

1. All of the above
2. Soak beans or whole grains in whey over night to break down phytic acid and other hard-to-digest proteins.
3. Use to inoculate sauerkraut and other fermented vegetables.
4. Use to make kvass (recipe above).
5. Use to make naturally carbonated lacto-fermented sodas *(recipes on 50)*.

Kvass

1 c whey
1 red beet, sliced into small 1/4" cubes
1 t salt
Non-chlorinated water to fill

1. Put whey, beets, and salt into a quart jar
2. Top with non-chlorinated water.
3. Let sit for 2+ days at a room temperature of 72° or below.
4. When it starts to bubble and smell tangy, refrigerate and drink. The beets in the jar are also fermented and may be eaten.

whey soda fermenting

Lacto-fermented sodas, also called whey sodas, are quick and easy to make and are a tasty, healthy alternative to canned soda. They offer a similar sweet carbonated experience, yet are packed with as many pro-biotic micro-organisms as a serving of yogurt. The micro-organism do need a caloric sweetener to work their magic and proliferate. If you like a less sweet product, allow the drink to ferment longer--the lacto-bacilli will consume the sugars and transform them to lactic acid. While extremely beneficial and anti-oxidant, these drinks are acidic and can be hard on the teeth. If serving to children, make sure they rinse with water after consuming.

Whey Soda with Fruit

1/2-3/4 c whey
1/2-3/4 c sugar
1-2 c fresh fruit diced or 1 – 2 c juice. For soft fruit like berries, no processing necessary, just add to the jar.
Pinch of salt
Non-chlorinated water to fill

1. Place all ingredients into a gallon jar and top with non-chlorinated water.
2. Cap and shake well to dissolve sugar.
3. Loosen the cap and let sit in a warm spot (up to 90°) for 4-7 days.
4. Check daily. Once it starts to bubble it is done.
5. Bottle and refrigerate.

Whey Soda with Herbal Infusions

6 cups herbal infusion such as peppermint, hibiscus, lemongrass
1/2-3/4 c whey
1/2-3/4 c sugar

1. Make the infusion by steeping the herbs in non-chlorinated boiling water.
2. Let the tea cool to 90°.
3. Add the sugar and the whey and mix to dissolve the sugar.
4. Cap loosely and let sit in a warm spot (up to 90°) for 4-7 days.
5. Check daily. Once it starts to bubble it is done.
6. Bottle and refrigerate.

** These drinks are strongly carbonated. They are a living food and will continue to ferment. If you will not drink right away, leave the cap of the bottle or jar cracked open a bit so the carbonation can escape. Exploding bottles can be very dangerous!*

THE STORY OF HARD CHEESE

Or Why Hard Cheese is Hard

As I mentioned, we live in an unnatural (albeit glorious) state of cheese in which we can walk into any good grocery and choose from as many as two hundred different cheeses. But it didn't used to be that way—it used to be that cheese varieties were regional and you'd have six to eight different styles to choose from—only those which were easily made with the naturally occurring local lactobacilli in the local climate.

Making a hard cheese is not so difficult, but making a specific hard cheese that resembles a cheese of that same name from the market is no small feat. Try the same cheese recipe five times and it will likely come out five different ways. This is because there are many variables that affect the outcome, some of which are tough to control in a home environment. Let's look at the process, step-by-step and learn about what's involved in making hard aged cheeses such as Cheddar, Jack, and Gouda.

Step One: Inoculate, Ripen, and Coagulate

The first steps in making hard cheese are very much like the feta cheese you just learned to make. The main difference is that slight variations of temperature or time in the beginning multiply into the aging process and impact how the cheese turns out. So it is important to be exact and to record your experiments (*see Home Cheesemaking Record Form*). All aged cheese starts by warming the milk to a target temperature and inoculating with culture. The milk will then need to ripen for a specific amount of time to allow the cultures to proliferate and acidify the milk. Different amounts of time will affect how firm a curd and the flavor of the cheese. Then the rennet is stirred in. The amount of rennet will also be specific to the cheese as will the amount of time needed to achieve a clean break. At this point the cheesemaking process will start to diversify.

Step Two: Cutting the Curds

After you get a clean break, the curds are cut. The size of the curds affects how hard the cheese will be. The size can range from rice grain size curds for a cheese like Manchego up to 1" cubes for Mozzarella.

Step Three: Cooking the Curds

Cooking the curds can be daunting, especially for the beginner. It requires that you increase the temperature of the curds and whey in your pot no more than two degrees every five minutes, stirring constantly until you reach a target temperature. For most cheeses, you'd be starting in the high 80s and heating to somewhere in the low 100s. We are talking tiny increments of temperature over a long period of time—sometimes as much as 45 minutes to an hour of paying attention and stirring. And so, you might ask, how am I going to do that? For the home cheese maker there are a few options.

The method that is most often recommended is the water-bath method. Set your pot of curds and whey into a larger pot in the sink. Place a thermometer in each pot. Use the tap water to set the temperature in the water bath ten degrees higher than the temperature you'd like the curds. Stirring constantly continue to adjust the temperature in the water-bath with the tap water, raising that temperature two degrees every five minutes.

If this sounds too fussy, you can try cooking the curds on the stove. Set your pot over the very lowest flame possible, such that the lightest puff of air would blow it out. If it still raises the temperature too quickly, you can turn it off and on or use a flame arrester between the pot and the flame to slow it down. In an act of great forgiveness, the universe gives us the simple gift that once you start working with larger batches of milk, say 16-20 quarts, to make hard cheeses (a good idea, as it is a lot of effort), raising the temperature slowly becomes easier.

But why, you might ask, does it have to go so slow? Can't we just bring it right up to the target temperature? The answer is no. The curds are delicate and still quite moist and gelatinous. In order to press them to achieve a hard cheese, they must be toughened up by expelling the whey. If you raise the temperature too quickly a tough crust will form on the outside of the curd, sealing the moisture in and preventing the whey from releasing. The result will be curds of an uneven consistency—tough on the outside, gelatinous on the inside. When

pressed together in a cheese press, these curds do not hold together well and the end result will be a crumbly chalky texture.

Cooking the curds could also be called "expelling the whey." We have already done this with our Feta. Even though we did not raise the temperature in the pot, with over 20 minutes of stirring in the pool of arm whey, the feta curds shrank and toughened up.

Step Four: Draining, Milling, & Pressing

Once the curds are cooked and the whey expelled, they are poured or ladled into a cheesecloth-lined colander and hung to drain. The curds mat together. After the correct amount of draining time, the curds are removed from the cloth and broken up by hand or with a fork. This is called milling. It is at this point that salt is usually added.

Then it is time to press the cheese. A cheese press is lined with cheesecloth. The curds are packed into the press and the cheesecloth is folded over the top of the curds neatly. The cheese is pressed at a low pressure for 15 minutes on each side before being redressed with a fresh cloth and pressed at the target pressure for a longer amount of time.

A cheese press is a pretty significant investment and is probably the very last piece of equipment you should buy. The press pictured is sold by Hoegger Dairy Supply and currently costs about $130. The pressure gage is another $60. For many cheeses a simple plastic mold with a follower that can be weighted with a gallon of water or books will suffice.

Cheese press with weighted gage assembled

Step Five: Aging the Cheese

After the cheese is pressed, the cheesecloth is carefully unwrapped from the cheese, and it is then air-dried for several hours or several days. The next step, aging the cheese, is probably

Cheese press disassembled (from top left, clockwise): Stainless steel ring, follower,washers and nuts, pressure gage, base

the most difficult for the home cheese maker, next to cooking the curds. This is because most cheeses need to age at a temperature much warmer than you can set your refrigerator and much cooler than room temperature and at a specific and high humidity.

In terms of reaching your target temperature and humidity (about 53 degrees and 85% humidity for many cheeses) the best option would be a cave. In France. If you live in a region where root cellars are still common, lucky you! You have a natural aging environment. But if you don't have the good fortune to have a root cellar, your options include buying a dedicated refrigerator and controlling it with a separate thermostat or purchasing a wine refrigerator. Better wine refrigerators will allow you to set both temperature and humidity and very high-end versions will allow you to set different environments in several different compartments so you can do more than one cheese at a time.

Even if you have been extremely careful with sterility, the likelihood that your cheese will be attacked by other bacteria or mold during the aging process is pretty high. Other home cheesemaking books suggest wiping your cheeses with brine or vinegar on a regular basis, but this cheese maker has found this to be a fairly thankless and unsuccessful task. Covering the cheese with a cheese wax is somewhat more effective at protecting the cheese from unwanted microbes.

Some of the authors hard cheeses, Gomano (see recipe page 80), Leerdammer and Gouda in cheese wax below.

So how did the farmwives do it back then? And how do the big dairies do it now?

Traditional systems for cooking curds and aging cheese were worked out over many generations and were based on terroir and bio-regional seasonal cycles. Modern cheese dairies have stainless steel rooms that can be efficiently sterilized, computer-controlled water baths for cooking the curd and multiple computer regulated refrigerators for aging the cheeses.

Thus goes the story of hard cheese. If it seems like a lot of work, it is. I tell you all this not to discourage you, but to give you an even deeper appreciation for the art, science and time that goes into creating all those cheeses that can be found in any grocery store. If you want to make hard cheese in your home it is not impossible. But it will require work and ingenuity to create your own "terroir" based on your own kitchen, and the ambient temperature of your area.

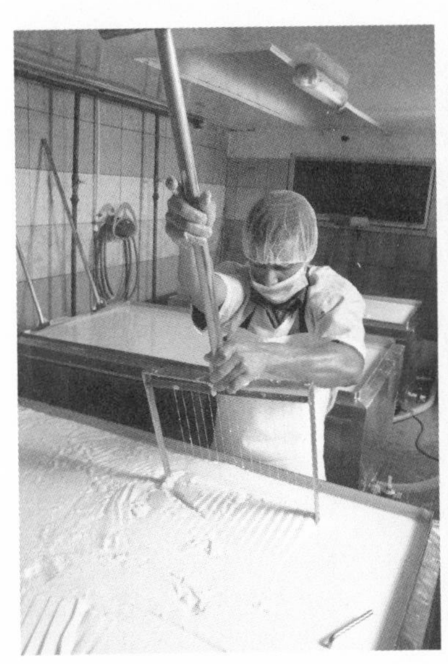

Cutting the curds in a small scale commercial dairy

A commercial cheese cave

India: Panir

The process for making Panir is very similar to Ricotta. The resulting curd is dry and readily soaks up the sauces and flavors around it.

1 gallon whole milk
3/4 c lemon juice
Near boiling water *(optional)*
Warm tap water

1. Over medium high heat, bring milk to a gentle boil, stirring constantly towards the end.
2. As the milk starts to foam up, quickly reduce the heat and pour in the lemon juice.
3. Stir for 10-15 seconds until curds form.
4. For a softer Panir, stir in 1-2 c near boiling water as soon as your curds form.
5. Allow curds to settle in the pot 5 minutes
6. Pour into cheesecloth lined colander. Rinse curds by holding under lukewarm running water for 10-15 seconds. Knot ends and hang to drain.
7. Hang for 2 hours or press in a small cheese press under about 5 pounds of pressure.
8. Slice and use in your Indian dishes. Keeps for two weeks in the fridge.

France: Chevre & Her Sisters

Chevre is traditionally made from goat milk, however there are many cow milk cheeses made by the same process. Two of these are Neufchatel and Gervais, which are made from cow milk plus heavy cream. All of these cheeses use a long ripening period and small amounts of rennet to achieve a delicate, richly flavored, moist curd. Like the Lemon Ricotta they can all be flavored in a multitude of ways, savory to sweet, and be used for cheese-based desserts.

Chevre

1 gallon goat milk

1/4 c cultured buttermilk or 1/8 t freeze-dried mesophilic culture

2T diluted liquid rennet *(Put 3 drops of liquid rennet into 1/3 c cool water. Use 2T of this mixture.)*

1. Put the milk into the pot and heat milk to 72°-80° to bring the milk to room temperature.
2. Add buttermilk or culture and rennet and stir.
3. Let sit at room temperature for 8-12 hours until it is the consistency of yogurt. I have found it to work well to let it ripen a bit longer until the curds sink below the whey.
4. Ladle the curds into a colander lined with a fine cheesecloth and hang 6-8 hours, or until drained.
5. Mill the cheese with a fork to achieve an even texture.
6. Season and eat immediately or store for up to 10 days in the refrigerator. This cheese can also be frozen for later use, up to 6 months. I often package in 8 oz portions in freezer paper—the perfect size for a dinner party or potluck.

Neufchatel

1 gallon whole cow milk

1 pint heavy whipping cream

1/4 c buttermilk or 1/8 t mesophilic culture

1T diluted rennet mixture

(3 drops in 1/3 C non-chlorinated water)

Follow the same steps as for Chevre. This cheese may be drained up to 12 hours and then pressed under the weight of two bricks in a cheese form or colander in the fridge for 12 hours before being milled, flavored, and shaped.

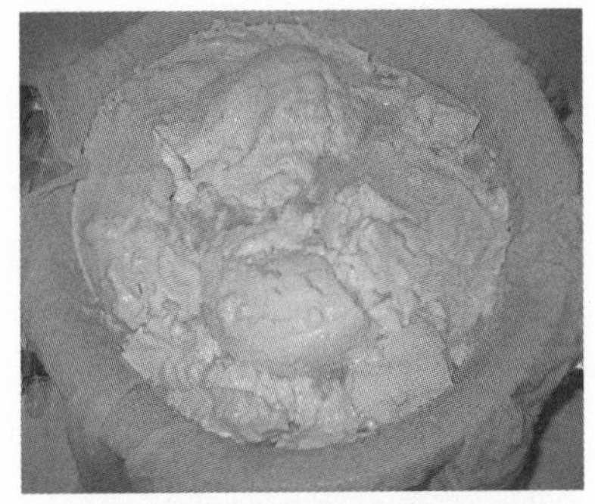

Chevre curd in a cheesecloth-lined colander and ready to hang and drain

Flavoring Fresh Cheeses

Add any of the following to 1 lb. of Chevre, Lemon Ricotta, Neufchatel, Gervais, or Yogurt Cheese. Fresh herbs and garlic are always best, but dried, or powdered may be used. Measurements are approximate. It is always best to season "to taste," and you have absolute permission to taste as often as you like! Once you have mixed ingredients thoroughly, shape into logs or balls, and refrigerate. Any non-iodized salt may be used.

Chives. 1t salt (or to taste), 1T garlic, 2T chives.

Dill. 1t salt, 3t minced dill.

French Onion. 1t salt, 3T dried minced onion. Roll balls or logs in cracked black pepper *(optional).*

Herbs de Provence. 1t salt, 2-4 gloves pressed garlic, 3-4T minced fresh tarragon, basil, rosemary, marjorum, oregano, and parsley.

Pineapple Walnut. 1t salt, 1/2 c chopped walnuts, 3/4 c crushed pineapple.

Sesame. 1 tsp salt, 1/2 t garlic, 2T minced onion, 1t paprika, 2t tamari, a few drops of sesame oil. Shape and roll in sesame seeds.

Tomato & Chili. 1t salt, 1t garlic, 1T minced onion, 1t chili powder, 6 oz. tomato paste, 1/4 t cayenne, chopped jalapeño, paprika or hot sauce to taste.

Other possibilities: Horseradish, sweeten with chopped strawberries or figs, blend with smoked salmon.

Gervais

2-2/3 c whole milk
1-1/3 c heavy cream
1 drop liquid rennet in
2T cool water

1. Heat to 65°.
2. Stir in rennet.
3. Let sit for 24 hours or until coagulated.
4. Drain in a cheesecloth 4-6 hours.
5. Mill, flavor and eat.

Mexico: Queso Fresco

Queso Fresco (fresh cheese) is a mild, fresh Latin American farmer's cheese. Queso Fresco recipes usually call for mesophilic culture, with or without an addition of vinegar, though in some regions Queso Fresco may be an acid cheese. In this recipe you will try your hand at cooking the curds and pressing the cheese.

1 gallon milk
1/4 c buttermilk or 1/8 t freeze-dried mesophilic culture
1/2 t liquid rennet diluted in 1/4 c water, or 1/16 t powdered rennet
1T salt

1. Put the milk into the pot and heat milk over medium heat to 90°.

2. Add culture and stir in thoroughly with a figure eight motion.

3. Add rennet and stir in thoroughly with a figure eight motion.

4. Allow to set, maintaining the temperature at 90° for 20-45 minutes until you get a clean break.

5. Cut curds into 1/4" pieces.

6. Over the next 20 minutes gradually increase temperature to 98° stirring gently. This can be done in a hot water bath, or over a very low flame.

7. Pour into colander (no cheesecloth) to drain off most of the whey and return the curds to the pot.

8. Keeping the curds warm in a water bath or pilot lit oven, stir in 1 t salt every 5 minutes for a total of 3t and 15 minutes furthe ripening time.

9. Line cheese press or 5" cheese mold with cheesecloth, ladle the curds into the mold, place the follower on top and press 3-6 hours under light pressure.

Finished Queso Fresco

Rob Reeves

Filling the cheese press. The press is first lined with cheese cloth and the curds are spooned in and tapped evenly into place. In this picture a standard weighted cheese press is used, however a plastic tomme style cheese press will work equally well. Simply weight with a heavy book or a brick.

Italy: Mozzarella, The Long and The Short of It

Mozzarella belongs to a class of cheeses called "pasta filata" or stretched curd cheeses. Other cheeses in this category include Provolone and Romano. The acidification of the curds allows the cheeses to melt and stretch. The curds are formed into logs or balls or even braided. Stretching lines up the protein strands form the strings of string cheese.

Offered here are two recipes. The shorter form uses citric acid to get a sure and quick acidification. It is not a perfect recipe, for reasons we'll talk about below, but it does work and produces a good quality chewy mozzarella that can be eaten by the slice or grated and melted on pizza or in lasagna. It should not be confused with the "30 Minute Mozzarella" popularized by cheesemaking supply houses, which by-passes many steps and produces a cheese that is rubbery and doesn't store well.

Traditional long form Mozzarella uses thermophilic culture alone to acidify the milk and may take up to four hours to reach the stage where the cheese stretches properly. Added to this is the issue that seasonal milk differences may make it such that the milk will never reach proper acidification without the addition of a small amount of citric acid.

Both forms of Mozzarella will require heat resistant gloves for stretching the mozzarella. I use a double glove system, with thin cotton cloves inside for insulation and rubber gloves outside to keep your hands dry. White cotton gloves for handling artwork are readily available, but any thin cloth gloves will work. Heat resistant silicon gloves are generally too thick for the amount of finger dexterity required and other heat resistant gloves are woven and would not work handling a liquid.

New World Mozzarella

1 gallon milk
1-1/4 t citric acid powder
1/4 t liquid rennet *or* 1/16 tsp powdered rennet
Brine: 1/2 c salt in 1 quart water
Heat resistant gloves

1. Put the milk into the pot.

2. Dissolve citric acid in 1/4 c cool water and mix thoroughly into milk.

3. Over medium heat, bring the temperature to 88°. The milk will already start to curdle lightly. This is what makes this recipe imperfect, but don't worry. It is going to come out fine!

4. Add the rennet and stir briefly. The milk is continuing to curdle before your eyes, and you do not want to break up the curds.

5. Let sit for 15 minutes. You won't exactly get a clean break, but it will be obvious that the curds have set.

6. Cut the curds into 1" cubes.

7. Slowly bring the temperature up about one degree every minute, stirring until you reach 108°. You can do this over a low flame and it should take about 20 minutes.

Stretching the Mozzarella

8. Drain into colander (no cheesecloth).

9. Return the whey to the pot and heat on medium-high to 155°-160°.

10. While it is heating, mix the brine in a medium sized bowl.

11. When the whey reaches temperature, place all or part of the curds into the hot whey (depending if you want one large ball or more smaller balls). Wait a moment or two for the curds to soften. Pull out the mass of curds and press them together into a ball. Then start to stretch and fold. On the first pass the cheese will be uneven and break apart as it stretches. Continue to pull and fold until the consistency is smooth and the cheese stretches easily. Do not over stretch! This will make the cheese tough. Once it is stretching easily and is a smooth consistency, shape into a ball and place into the brine.

12. Leave in brine for 30 minutes.

13. Remove and pat dry.

14. Eat immediately or wrap in plastic and refrigerate. The next day the cheese may be sliced or grated for use in pizza, lasagna, or sandwiches.

Wrap the stretched cheese around the thumb to form a a ball. It can also be braided, or folded and pressed into logs for a string cheese effect

Old World Mozzarella

2 gallons goat or cow milk
1/8 t themophilic culture or 1/2 c yogurt
1/2 t liquid rennet mixed into 1/4 c water or 1/16 t dry rennet

1. Put the milk into the pot and heat over medium heat stirring in a figure 8 until you reach 90°
2. Add culture and mix thoroughly.
3. Add rennet and mix thoroughly.
4. Maintain the temperature and let ripen for 45 -75 minutes or until you get a clean break.
5. Cut into 1" cubes.
6. Cook the curds, bringing the temperature up no more than 2° every 5 minutes until you reach 100°
7. Leaving the curds in the whey, maintain the temperature at 100° in a hot water bath, or in an oven with a pilot light for 2.5 hours.
8. Test the curds for stretch by placing about 1/2 t of curds into near boiling water. After a moment scoop it out and try stretching. If it is not ready, it will break, rather than stretch.
9. If not ready, continue to ripen at 100° and test for stretch every half hour until you get there.
10. Once you get a stretch you can either continue to step 11, or refrigerate and stretch another day.
11. Continue as for short form Mozzarella, steps 9-14.

Yes, I Can Tell You How To Make Burrata

Once your Mozzarella has been stretched, pull it out into a pancake. Place in your filling and fold the pancake around the filling, shaping the cheese. Fillings vary. Sometime it is small pieces of Mozzarella combined with Crème Fraîche or Cream Cheese, sometime just Crème Fraîche or Cream Cheese alone. Be creative! For homemade Crème Fraîche and Cream Cheese, see page 77.

Troubleshooting Mozzarella

Problem: The curd does not stretch but breaks instead

Cause: Maybe the curds are fine. It sometimes takes a few rounds of simply stretching a little and folding before the curds will actually stretch.

Cause: Curds have not been acidified long enough. Return to pot to ripen longer, or refrigerate over night and try again.

Cause: Milk is at a time of year that it will never get acidic enough without the addition of citric acid. Unfortunately, this cheese will never become Mozzarella. Chop or blend the curd and eat as is. If you are starting over with the same milk, add 1 tsp of vinegar or 1/2 t citric acid in addition to your culture at the beginning.

Cause: Curd has gotten over-acidified. This can happen from time to time. The cheese will be brittle and break apart, even after a bit of gentle folding and stretching. Give up, eat the curds as a gunky mass and try again.

Greece: Halloumi

Originating in Cypress, Halloumi was traditionally made from sheep milk or a combination of sheep and goat milk. It is one step more complex than Feta and has a nice chewy texture.

1 gallon cow milk, goat milk, or mixture
1/2 t liquid rennet or 1/16 t powdered rennet
salt

1. Put the milk into the pot and heat over medium heat stirring in a figure 8 until you reach 90°. Turn off heat.
2. Add rennet and mix.
3. Cover and maintain temperature for 30-45 minutes until you get a clean break.
4. Cut the curds into 3/4" cubes and let sit for 5 minutes.
5. Over low heat, stirring slowly with the flat of your spoon, raise the temperature about 2° per minute until you reach 104°. This should take about 15 minutes.
6. Let sit for 5 minutes.
7. Drain the curds into a colander lined with cheesecloth and let drain 15 minutes. Reserve whey in the refrigerator.
8. Line a 5" mold or cheese press with a cheesecloth and spoon the curds into it. Fold the cheesecloth over the

curds, set follower on top and press at 8 pounds for 3 hours. 1 gallon of water in a jug will give you the 8 pounds of pressure you need. A large book will also work and may balance better on the mold.

9. Remove the cheese from the mold. If you wish, you can trim it into a square or slice horizontally to make two slabs.

10. Heat the reserved whey to 190°. Place the cheese into the hot whey, maintaining the temperature at 190° for 30-35 minutes.

11. Using a skimmer or slotted spoon, remove the cheese and place on a draining rack to cool, flipping once or twice until the surface is dry to the touch. It will take about 45 minutes to one hour.

12. Make a brine using 1-1-1/2 T salt per quart. Place cheese in a container and cover with brine. This cheese is ready to eat within 5 days and will keep in the refrigerator up to two months.

∽ DEEP CULTURE ∾

Direct Set, Freeze-Dried DVI and Reculturable Cultures

Now that you have a sense of basic cheesemaking it is time to get a deeper understanding of the cultures used. First it is important to understand the different forms the cultures are sold in.

Direct set cultures are an all-in-one, pre-measured dose of culture and rennet, with some maltodextrin added. They are often blended and sold as Buttermilk, Crème Fraîche, Chevre or other cultures even though these are all mesophilic culture.

Freeze-dried direct vat inoculation (DVI) cultures wake up almost instantly when they hit warm milk and can be stored indefinitely in the freezer. They are the most convenient and offer the most flexibility in their usage for occasional cheese making.

Reculturable cultures are meant to be cultured in milk to wake them up. They are then stored in milk and "fed" regularly when cheese is made. This was how cultures were traditionally kept going and passed around. Using reculturable cultures requires regular use and feeding and the knowledge or acceptance that eventually they will become contaminated with the local bacteria present in your kitchen.

Characteristics of Culture Communities

Cheese cultures are communities of micro-organisms, that work together. The different blends of organisms will create slight differences in productivity, flavor, and mouth-feel. Because there are many other variables we can't control efficiently in the home environment, the difference between one culture and another will be negligible for home cheese makers.

Still, it is interesting to note how the cultures are supposed to work. The chart below shows four variables professional cheese makers look at when considering which culture to use for a given cheese. Below that, we look at some of the specific blends on the market today.

Acidification: (LoA, MidA or HiA) How much lactic acid the culture produces.

Proteolysis: (LoPT, MidPT or HiPT) A process contributing to flavor development in aging.
Diacetyl: (LoD MidD or HiD) Fermentation compound that adds a buttery texture and aroma.
Gas: (loG, Mid G or HiG) Amount of CO^2 produced by the culture.

Mesophilic Starter Culture—Buttermilk

All freeze-dried mesophilic cultures or buttermilk can be used in any recipe that calls for Mesophilic culture, but will impart subtly different mouth-feel and flavor profiles. As well they are classified as low, moderate or high acidifiers. They are primarily blends of 3 organisms: lactococcus lactis ssp. lactis (LL), lactococcus lactis ssp. cremoris (LLC) and lactococcus lactis ssp. biovar diacetylactis (LLD), all of which perform well in the 72°-100° range. Different suppliers have their own blends and this list is not exhaustive, however here are some you might come across:
Meso I: LoA. Cheddar, Jack
Meso II: Mid-HiA, LoG, LoD. Brick, Brie, Colby, and stretched curd cheeses.
Meso III: Mid-HiA, LoG, LoD, HiP. Clean flavor, closed texture. Edam, Gouda, Havarti
MM series: MidA, MidG, HiD. Creamy open texture. Brie, Blue, Edam, Chevre
MA Series: Mid-HiA, LoG, LoD, HiPT. Aged hard cheeses. Cheddar, Jack, Stilton, Colby
Aroma B and Flor Danica: MidA, MidG, HiD. Cream cheese, Sour Cream, Cultured Butter

Thermophilic Starter Culture—Yogurt

Thermophilic cultures are used to make hard, aged Italian and Swiss cheeses like Parmesan, Romano, Provolone, Mozzarella, and Emmental/Swiss. Thermophilic cultures may be used to add a milder flavor to your Camembert/Brie and allow for longer aging. Most Thermophilic cultures blend streptococcus themophilius with lactobacillus delbreukii ssp. bulgaricus or lactococcus helveticus. Thermophilic cultures perform best in the 90°-115° range.
Thermo B: MidA, lo-MidD, MidG. With bulgaricus. Mozzarella, Parmesan, Romano, Provolone, other Italian cheeses.

Thermo C: MidA, lo-MidD, with helveticus. Mozzarella, Parmesan, Romano, Provolone, other Italian cheeses.

Secondary Cultures

Used in conjunction with your Mesophillic or Thermophilic cultures to create specific results.

Penicillium Candidum: White mold. Produces the fuzzy white surface in "bloomy rind" cheeses including Camembert and Brie as well as French goat cheeses. Several strains available produce flavors mild to strong. ABL, HP6, SAM3, VS

Geotrichum Candidum: Helps with rind formation and protects against unwanted molds when making Camembert and Brie. Strains available include—Geo 15 mild, Geo 13 medium, Geo17 strong.

Penicillium Roqueforti: Blue Mold. Creates the characteristic blue veining and adds a strong, clear blue cheese favor. Available in several strains. For making blue cheese including Stilton, Collummiers, Gorgonzola.

Bacteria Linens: Red Mold. Requires high pH (low acidity). Used to create red or orange rinds in cheeses such as Brick, Limburger, and Muenster.

Proprioni Shermanii: The Gassy Culture. For the maturation and eye formation of Swiss type cheeses, such as Gruyere, Emmental, Gruyere, Jarlsberg, Compte, Tilsit, and Appenzeller. Gas production in the ripening process creates the characteristic holes in these Swiss-style cheeses and the culture contributes greatly to the taste and aroma of these types of cheeses.

∽ MOLD RIPENED CHEESES ∾

After months of struggling to cook curds and successfully make Manchego, I threw my hands up and went back to my research. I wanted to make something "fancy." Something I could bring to a party that would really wow people. I assumed that Brie, my favorite all-time decadent cheese would be too difficult. But when finally got into it, I discovered it required no cooking of curds and no special ripening environment. It could be ripened in the refrigerator. Part of the process could even be done at room temperature, if it were below 65°. That was no problem--sounds like summer in the San Francisco Bay Area where I live. So I gave it a shot and it came out with perfect results the very first time. Since then, I have had some failures, and I always taste this cheese before I serve it, as it is occasionaly colonized by something unpalatable, but overall I find this cheese is quite doable and 90% reliable as a fancy homemade cheese.

Additional Equipment for Mold Ripened Cheeses

In addition to the equipment already mentioned, for mold-ripened cheese you will need:

Two Camembert molds. These are cylinders 4 inch internal diameter, 4.5 inches high with small holes spaced for drainage. Two molds will accommodate a one gallon batch of Camembert/Petite Brie.

Two-Ply Cheese Paper. This looks like plastic on the outside and wax paper on the inside, but it is actually a natural breathable cellophane. The cheese must breath or it will rot! It must not dry out or it will not ripen properly, I have not found a cheap substitute from the grocer, but the paper is inexpensive and readily available from cheese supply houses. It is called, simply, "cheese paper" and comes in several sizes.

Camembert/Petite Brie

This classy bloomy rind cheese is not difficult to make and can be aged primarily in a refrigerator with some success. This recipe will make two four-inch rounds of cheese. You can make a true Brie by using a 12 inch ring and starting with 2-3 gallons of milk. This recipe calls for two white mold cultures. If you do not want to order them, you can "borrow" the culture from some store bought Brie. Simply take about a tablespoon of the cheese, including a piece of the rind, blend it in a blender, with 1/4 c milk. Strain this and add the now inoculated milk to your pot.

2 gallon pot with lid
Thermometer
Curd knife
Slotted spoon
Camembert molds
Cheese mat (2)
Cheese draining rack (2)
Cheese paper (2)
Wax paper
Aging container

1 gallon whole milk
1 pint cream (for double cream brie)
1/8 t mesophilic starter
1/16 t penicillium candidum
tiny amount geotrichum candidum *(optional)*
1/16 t dry rennet or 1/2 t liquid rennet in 1/4 c water
salt

1. ***Ripen and Coagulate.*** Heat the milk to 90° and then turn off the heat. Stir in all three cultures. Maintain temperature for 90 minutes. Stir in the rennet. Let coagulate 60 minutes or until you get a clean break. A second method for this step is to heat the milk and add the culture and rennet all at once at the beginning, then ripen for 2-3 hours, until you get a clean break. This yields a milder cheese.

2. ***Cut and cook the curds.*** Once you get a clean break, cut the curds into 1/2 inch pieces. Stir gently with the flat of the spoon for 10 -15 minutes, cutting any large pieces against the pot with the edge of the spoon. If the temperature has dropped, slowly raise it back to 90° as you continue to stir. Let the curds settle for 5 minutes.

3. ***Drain.*** Sanitize and set up the cheese forms, sandwiching the two Camembert forms between the draining racks and mats. Insert a sheet of wax paper, then the forms. Once you ladle in the curds, set another sheet of wax paper on top, then a mat, then a draining rack. For now, assemble those off to the side. Pour off some whey to expose the curds. Use a slotted spoon to ladle the curd into cheese forms, alternating between the two forms to distribute the curd evenly. Holding the sandwiched forms firmly between two hands, flip every 15 minutes in the first hour, then once per hour for 5-6 hours.

Camembert is one of the fancier cheeses that is do-able for the home cheesemaker.

Let sir overnight (optional). Put in the fridge for 1-3 hours to firm *(optional)*.

4. **Age.** Pull off the Camembert forms, leaving the cheeses on their draining platform. Salt the cheeses, thoroughly sprinkling all sides. Set the cheeses on their mat and rack into the aging container. The aging process has two parts. The first part can be done at room temperature if room temperature is no more than 65°, if this is not possible you can do it in the refrigerator. It is important that the cheese does not dry out, as moisture is required for the mold to grow. It is especially important to cover the aging container if you use the refrigerator. Another method for this part of the process is to create your own mini cheese cave using a camping cooler and two blue ice packs. Keep a thermometer inside the cooler to monitor the temperature and switch out the ice packs every twelve hours or when the temperature rises to about 60°. Flip the cheeses daily and drain moisture until white mold appears. After two or three days, remove the draining mat, leaving the rack so that the mold does not form around the mat and stick to it. The mold will bloom in 5-7 days at room temperature or 10-14 days in the refrigerator.

Once you see mold, wrap the cheese in cheese paper. Ideal ripening temperature is 52°-55°, however it will also work on the warm setting of your regular refrigerator (about 42-43°). Age 4-6 weeks. The cheese ripens from the outside to the center. You can test for doneness by squeezing the center to see if it is evenly soft. You can also cut your cheese open to check, but once it is open you cannot return to aging it. When the cheese is ripe, it should be eaten within 7-10 days. After that the quality starts to go down and the cheese will either dry up or turn into a smelly goo.

The two camember forms are sandwiched between the draining racks and the draining mats.

Rob Reeves

Ladling the camembert curds into the forms in the sink, alternating between each form so the curds are distributed evenly.

The Camembert form filled with curds. Depending on your milk, there may be more curds than room in the form. If this is the case, wait a few minutes for the curds to drain and continue filling.

The Camembert in the forms after draining overnight.

The Camembert cheeses with the molds removed after draning over night

White mold blooming on the cheese

Camembert rind fail: The mold has formed around the draining mat and has been pulled away from the cheese when the mat was removed. To avoid this, remove the draining mats after 2-3 days of ripening, once the cheese has firmed up, but the mold has not yet formed.

Troubleshooting Camembert

Problem: Cheese turns yellow & brittle in the aging process and mold does not form.

Cause: Aging environment is too dry. This cheese probably cannot be rescued, in the future cover the aging container with a plastic lid.

Problem: Cheese is overripe and runny just inside the rind and hard towards the middle.

Cause: Too much moisture. Open up the ripening container a bit more during the initial blooming process.

Cause: Not aged long enough. The moisture will redistribute in the aging process. Try aging longer next batch.

Blue Cheese

Blue cheese runs creamy to dry, but they all have the characteristic blue cheese flavor which is achieved with Penicillium Roquforti. As with Brie, if you do not want to order the culture, you can "borrow" it from some store bought blue cheese. Simply take about a tablespoon of the cheese, including some of the blue veining. Blend it in a blender, with 1/4 c milk. Strain this and add the now inoculated milk to your pot. For your enjoyment I offer two different blue cheeses recipes. Castle blue is the creamier version, while Stilton is the crumblier. The process of making Castle Blue is similar to the Petite Camembert we just made, except that it uses the blue mold and requires running the cheese through with a knitting needle once the mold has bloomed on the surface to create spaces for the blue veining. The crumblier Stilton is hung for a day and then drained and broken into pieces before being lightly pressed together. This creates the irregular spaces for the blue mold to develope.

Once you have mastered these two, go look up the recipe for Gorgonzola (not included here, but easy to find). Gorgonzola requires making two batches of the same cheese twelve hours apart and packing one batch inside the other.

Castle Blue

In additon to the equiptment used for Camembert you will need:

Awl or Knitting Needle, sterilized
One Camembert Mold
1 Large Sheet 2-ply Cheese Paper

1 gallon milk
1 c whipping cream
1/8 t mesophilic culture
a dusting of penicillium roqueforti *(see intro above)*
1/2 t liquid rennet or 1/16 t dry rennet

1. ***Ripen and coagulate.*** Put the milk and the cream into the pot and heat to 90°. Turn off the heat. Sprinkle mesophilic culture and mold onto the milk and top stir. Cover and let ripen for 1.5 hours, maintaining the temperature. Add rennet and top stir. Let coagulate 1 hour or until you get a clean break.

2. ***Cut and cook the curds.*** Cut the curds into 1" pieces. Let stand for 10 minutes. Stir curds gently for 30 minutes at 90° until they shrink in size and start to mat together. Let settle.

3. ***Drain.*** Pour off whey. Set up mold with mats and racks. Ladle the curds into the mold. Flip every 15 minutes for the first hour and every 2 hours after that, then let sit overnight. Remove from mold and salt all sides.

4. ***Age.*** Place on rack in ripening container and cover. Ripen between 50°- 60° and 90% humidity. If you do not have access to this temperature range, use the cooler and ice pack method mentioned in the Camembert recipe above. Humidity may be maintained by covering the cheese in the aging container with a plastic lid. Flip daily for one week, pouring off whey from the container. After one week poke 8-10 holes in the cheese with a sterile knitting needle or awl. Return to container. In ten days blue mold should be visible on outside of cheese. Pierce cheese again at two weeks. The rind will turn blue/grey and the cheese will start to soften. Wrap in parchment paper or cheese paper and store in the refrigerator for up to one month.

Stilton Blue

Same as for Camembert

1- 4" Camembert mold or one 5" cheese mold

1 large sheet of cheese paper

1 gallon cow or goat milk

1/8 t mesophilic culture

1/32-1/16 t Penicillium Roqueforti OR 1/4 c blended blue cheese culture

1/16 t dry rennet OR 1/2 t liquid dissolved in 1/4 c water

1T salt

1. ***Ripen and coagulate.*** Heat milk to 90°. Sprinkle mesophilic culture & Penicillium Roqueforti (or cheese blend) onto milk and stir. Maintain the temperature and let ripen for 90 minutes. Add rennet and stir. Hold temperature for 1 more hour or until you get a "clean break."

2. ***Cut and cook the curds.*** Cut the curds into 3/4 inch pieces. Maintain the temperature at 90° (heat if needed) and gently stir every 1 to 2 minutes for 30 minutes.

3. ***Drain.*** Pour off most of the whey and ladle curds into a cheesecloth lined colander. Tie ends, hang and drain overnight.

4. ***Salt and mill.*** With your hands, break the curd mass into small pieces 1/4 -1/2 inch. Add 1 T coarse salt and mix thoroughly

5. ***Press.*** Place the Camembert form onto the draining mat and spoon curds in. Use a clean spoon to press the curds into place. Put wax paper or saran wrap on top of the cheese and weight with a jar or something similar. The next day flip your mold and weight again. Repeat this for 3-4 days or until the cheese holds its shape.

6. ***Age.*** Place cheese on draining rack inside an aging container. Close the container—you want it moist without water beading and falling on the cheese. Let it sit at 50°-60° for about a week. Check the cheese daily, drain off the whey and flip the cheese. After a week you will see blue mold over the surface. Age in the container at around 42°-43°. It will work on the top shelf of your refrigerator. At around 30 days take a thin slice of your cheese and check for blue veining. Once you see veining, wrap in cheese paper and continue to age in the refrigerator. For a drier cheese salt the outside of the cheese 2-3 times in the first two weeks after the cheese is covered with blue mold. This will dry the surface and produce a firmer rind. Once you have a good rind, you can poke the cheese with a sterile skewer to continue veining the blue mold.

St. Maure and Other Moldy Shapes

There are a number of bloomy rind cheeses that start with Chevre or a chevre sister that bridge the gap between easy and fancy. They all are made by a very similar process, but are drained in different shaped molds. Some are dusted with ash and others left plain. Traditionally Valencay (don't forget to add the cedille) is a pyramid with the top lopped off while St. Maure is in a log shape, I usually use a small round mold to drain my cheeses. Whatever shape appeals to you, here's how:

1. Follow the recipe for Chevre or Neufchatel. When you add the mesophilic culture include 1/16 t penicillium candidum and a dusting of geotrichum candidum. Otherwise proceed as usual until the curd has coagulated.
2. Once coagulated, ladle the curd into two small draining forms and let drain over night. Flip out onto a draining matt over a draining rack. Salt lightly on all sides—this will be a delicate operation as the cheese is quite soft. Ripen as for Camembert. After the mold blooms, wrap in cheese paper and eat within 7-10 days.

A St. Maure style cheese that was dusted with ash before ripening.

Both white mold and blue mold are persistant and will permanently colonize your ripening environment. Pictured here is a blue cheese that has been colonized by white mold as well. Traditionally these cheese were aged in isolated ripening caves

This blue cheese shows both the natural blue mold veining from the irregular spaces between the curds and the straight veins from being run through by the cheesemaker

Sour Cream/Crème Fraîche

Add 1/4 c buttermilk or 1/8 t powdered mesophilic culture to 1 pint of cream in a jar or bowl. Cover and set at warm room temperature up to 90° until it clabbers. For Crème Fraîche refrigerate and use as soon as it sets. For Sour Cream, let ripen 2-6 hours longer before refrigerating.

Rob Reeves

A food processor churns butter quite efficiently.

Cream Cheese

Make as for Sour Cream above with the addition of two drops of liquid rennet, diluted in 1/4 c water. Once it coagulates, hang and drain to desired consistency.

Yogurt Cream

Add 1/4 c plain yogurt to 1 pint heavy whipping cream and mix well. Incubate overnight at 110°-120°. I usually make this in a pint jar, wrap the pint jar in a heating pad and hold it in place with a Camembert mold. The cream does not set into a gelled yogurt, but does thicken and develop a good flavor. Excellent lightly sweetened on fruit!

Rob Reeves

The butter after churning. You can see a clear separation of fats from liquids.

Rob Reeves

The butter and buttermilk are poured into a butter muslin lined colander over a pot to catch the buttermilk. Then the butter is massaged against the butter musline to separate it from the buttermilk.

Sour Cream Butter

1. Start by making Sour Cream as above.
2. Chill to 58°. Too warm and the fat will be soft and will not give a clean separation. Too cold and the fat will be too hard and challenging to drain.
3. Churn. Agitate until butterfat separates from buttermilk. You can do this in a food processor, in an old-fashioned butter churn or by shaking inside a jar. The food processor method takes about three minutes, hand churning or shaking can take up to twenty minutes.
4. Drain. When the butter separates, drain in a strainer lined with butter muslin. Use a spatula to scrape and mix butter to let all the buttermilk drain. Salt to taste 1/4-1/2 t per pint. Press into a small bowl or a butter form and chill.

Rob Reeves

The butter after draining and ready to press into a butter form or small bowl

Yogurt Cream Butter

Make Yogurt Cream as above. Follow steps 2-4 as for Sour Cream Butter above.

Buttermilk

The left over liquid from making butter is real cultured buttermilk, consisting of the liquid and proteins of the cream. You can use it to culture cheese, make buttermilk pancakes, or a delicious ranch dressing. Buttermilk will keep frozen for about a month for using as a culture. I like to freeze it in ice cube trays. Two cubes are enough to inoculate 1 gallon of milk.

Homemade Ranch Dressing

Reserve 1 heaping tablespoon of the sour cream in a jar before churning to turn into butter.

Add to this 1/4 c buttermilk (from making your butter), finely chopped chives and dill, salt, and pepper to taste. Shake well.

Rob Reeves (all small photos)

People who have dairy animals learn to make cheese and do so on a regular basis so as not to loose those precious proteins. In the process they adapt recipes for efficiency. Here are just a few cheese we have adapted locally.

Cheater Cottage Cheese

1. Proceed as if you are going to make feta, through to the clean break.
2. Cut the curds into 1/4" pieces.
3. Slowly cook the curds over a low flame, stirring constantly until you reach 102°. This should take 20 minutes, or a bit less than one degree per minute.
4. Pour curds into a cheese cloth lined colander.
5. Rinse curds in lukewarm water.
6. The curds will be quite "squeaky" in texture. Pack into a closed container and let sit in the refrigerator overnight. The next day, add 1/4 c of cultured buttermilk or heavy whipping cream. Salt or sweeten to taste if desired. Mix thoroughly.

Green Faerie Farm Gomano (Faux Goat Romano)

We make this cheese with goat's milk, but it could also be made with cow's milk. It can be eaten mild without aging as a sliceable meltable white cheese or it can be aged and dried in the refrigerator for a dry, grating cheese.

1. Heat two gallons of goats milk to 90°, stirring over medium heat.
2. Inoculate with 1/4 t freeze dried thermophilic culture or 1/2 c yogurt.
3. Hold the temperature and let ripen 1 hour.
4. Add 1/16 t powdered rennet or 1/2 t liquid rennet diluted in 1/4 c water and stir thoroughly.
5. Hold the temperature and let sit 1-2 hours until it has gone past the clean break stage. The curd will pull away from sides and float in a pool of whey.
6. Cut the curds into 1/2" chunks.
7. Slowly bring the temperature up to 105° over a half hour, stirring constantly.
8. Let the curds sit in the whey another 30 minutes.
9. Drain into a cheesecloth-lined colander and hang to drip a few hours.
10. Break up the curds and press into a cheese mold. Press well to remove air. Cover with a follower and press under 8-10 pounds for one hour. Then take out of the press, remove the cheese cloth, flip over, redress, and press for

12 hours at 8-10 pounds of pressure.

11. Remove the cheese from the form and place into a near-saturated whey brine (1/2 gallon of whey and 1 cup salt) for 12-20 hours.

12. Air dry three days at room temperature. The cheese may be eaten at this stage or aged for a drier cheese. To age, wrap in a cloth and place in the refrigerator. If mold forms on the cheese, wipe with a vinegar salt solution.

Ruby's Cheater Cambozola

Follow the recipe for Camembert given in Section 3. When half the curds are ladled into the forms, dust a small amount of blue cheese mold onto the curds. Ladle the remaining curds into the form. Age as for Camembert. Your cheese will be creamy like Camembert but have a fine blue cheese flavor.

Dairy Cheese Endnote: Failure As Opportunity

Despite our best effort, every so often something fails. There's the Chevre that didn't set, the Blue that never pulled together, the Camembert with a terrible off-flavor. Of course, you can always throw it in the compost if you are completely discouraged. However, there is almost always some creative way to rescue and utilize that protein, if you can get over the fact that your project in no way resembles what you were trying for.

If your curds never set, that protein, unless unpleasantly sour, can be used in smoothies or fed to animals. The food processor is your friend. Anything that has gotten through to the curd process can be drained and milled. Try adding a bit of milk, cream, or buttermilk to improve the texture and some salt and flavors.

Occasionally the cheese just tastes rotten. Aged cheeses especially offer opportunistic bacteria a free lunch. Off flavors usually can't be fixed, however there's often someone whose palate can bear it. If not, it returns to the earth and we have the chance to ponder what went wrong and start again. Use the Homemade Cheese Record Form (Appendix D) to keep track of all the variables and try, try again.

The author's refrigerator, showing finished Feta (left), Yogurt (back) and Camembert (stacked in the front), as well as a host of fermented veggies *(right)*.

Nuts E

& Seeds

Whether you are vegan, lactose intolerant, or eat a paleo diet, nut and seed based yogurts and cheeses are a great option to satisfy the human need for a lipid and protein rich fermented food. Many commercially available non-dairy products are highly processed and contain non-food additives. Making non-dairy cheese at home ensures that you know your ingredients and can vouch for the quality and wholeness of your food.

As compared to their dairy cousins, nut based cheeses are incredibly easy to make and consist primarily of finely blended nuts or seeds which are cultured for a short time, then drained, or pressed and flavored.

Because nuts have a high percentage of oil, which denatures when heated, the focus here is on projects that use raw nuts as their base, with few additives or stabilizers. Like their dairy counterparts, many of the recipes feature culturing with living bacteria for a resulting living food. There is a limit to what you can do with raw cultured nuts, so I included a few recipes that involve cooking or the addition of stabilizers or gelling agents. In the yogurt section I include soy and coconut options and because not every person who wishes to avoid dairy is vegan, I included the possibility of using gelatin as the stabilizer. If you are vegan or vegetarian look for unflavored vegan gelatin. If you are a non-dairy omnivore, high-quality bovine gelatin maybe used.

Equipment Specific to Nut Cheeses

High-speed Blender

A Vitamix, Blendtec, or similar high-speed blender is essential to making nut cheeses. A normal food processor or kitchen variety blender will work in a pinch, but you won't get a completely smooth consistency, your cheeses will be slightly grainy.

Purchased new, a high-speed blender can be quite expensive, but with a little research you can find them used, and the motors are virtually indestructible. If you go direct to the manufacturer's websites you can also find reconditioned machines under warranty for a deep discount off the full retail price. Another option is to share the cost with a group of like-minded friends. Beyond making nut cheeses, high-speed blenders are great for grinding fresh grains, making smoothies, milkshakes, and sauces.

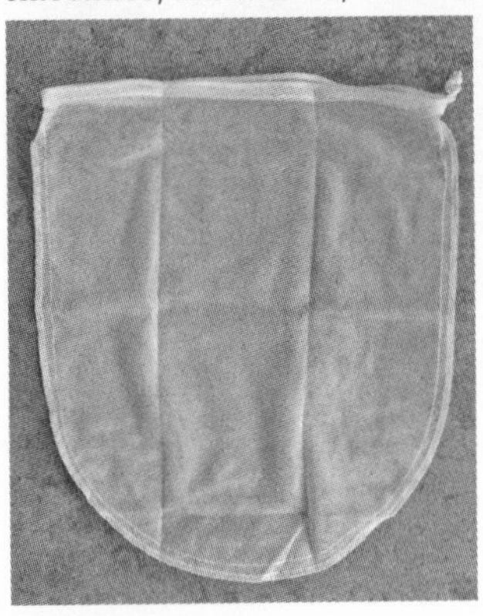

Nut Milk Bag

You can drain your nut milks and cheeses through a fine cheese cloth (not the flimsy store bought kind, but high-quality extra fine butter muslin from a cheese maker's supply house) or a muslin bag, but for the best results you may want to get a hold of a few dedicated nut milk bags. These are synthetic, have a very tight weave and drain efficiently. Nut milk bags are widely available online and may also be found at natural food stores. They are washable and reusable.

Cheese Press

A small 5" cheese press with a follower is useful for getting your nut cheeses into a cheese like shape, though in many cases a small bowl or ramekin will work just as well *(see picture page 53)*.

Dehydrator

For several projects a good quality dehydrator, or an oven with a dehydration setting is required. If you are investing in a dehydrator for the first time, I highly recommend one that has removable shelves and a temperature control.

About Nuts & Seeds

Nuts and seeds are high-density foods, rich in natural fats and proteins. The cheeses, yogurt, milks, and creams made from them should be eaten regularly, but in moderation (eating a cup of whole nut almond yogurt is equivalent to eating a full cup of almonds). Both nuts and seeds have a protective covering that is difficult for humans to digest. Soaking the nuts and culturing will help with this.

Most of the recipes here can be made with any nut or blend of nuts for a variety of interesting flavors and textures. Almond, macadamia, and cashew are the most versatile nuts for making cheese, but walnuts, pumpkin seed, brazil nuts, hazelnuts, pecans, pistachios, sunflower seeds, and pine nuts can all be used or blended. The addition of ground seeds such as flax, hemp, or sesame can also add flavor and texture. Don't be afraid to experiment, try different blends of nuts and seeds and develop your own recipes.

Nuts are protein packed flavor bombs that come with just the right amount of good fats to make them digestible.

Almonds are a fairly dry nut and even finely blended will give a somewhat grainy texture. They have a beautiful white color and a light refreshing flavor. Almonds should be soaked and skinned for best results. After a few hours of soaking the skins will slide right off.

Cashews are softer and moister than most other nuts and require less or little soaking to achieve a creamy consistency. They have a mild sweet flavor. The nut is actually the seed of the cashew apple. Cashews have lower fat and higher protein and carbohydrate content than other nuts.

Walnuts contain tannins, so cheeses made with walnuts will have a bitter edge.

Pumpkin Seeds have an exceptionally earthy flavor and make a beautiful dark green cheese.

Pine Nuts are especially creamy and fatty. Most pine nuts come from Italy and Spain, but some come from China. The Italian variety are more expensive, but have the best flavor. Of course if you live in a pine forest, you can also collect your own. Pine nuts have the most protein of any nut or seed.

Macadamia Nuts are quite fatty and like pine nuts will make a rich creamy cheese.

Hazelnuts Also called filberts when shelled, hazelnuts are much beloved for their distinct flavor. Hazelnuts should be soaked and skinned similarly to almonds.

Purchasing & Storing Nuts

While our recipes call for "raw" nuts it has become virtually impossible to purchase some kinds of nuts truly raw. Federal law mandates pasteurization of almonds, so that even almonds that say "raw" in the store are not. Pasteurization kills both the naturally occurring enzymes and the ability of the nut to sprout. If your nuts are organic, they are pasteurized by heat steaming. If they are not organic, they may have been fumigated with propylene oxide (PPO), a carcinogenic gas so nasty that it was banned by both the National Hot Rod and American Motorcycle Racing Associations, where it had been used as a fuel. Cashews and pistachios are not pasteurized, but they are usually heat steamed to crack open the shell.

For truly raw nuts, try buying direct from a farmer at a farmers market. Some specialty online stores also sell high-quality, truly raw nuts *(see Appendix A Resources)*.

Rejuvelac is is usually found with the other probiotic drinks in a natural foods market.

Store nuts in a cool, dry, dark environment to preserve freshness. The refrigerator is best, but a cool pantry will suffice. They can be stored in the bag they came in or in a sealed glass jar, such as a mason jar. Nuts stored this way should last for up to 6 months. Because of their high oil content Pine Nuts should be stored in an airtight container in the refrigerator or freezer.

Non-Dairy Cultures

You can ferment or culture your vegan cheese with a wide variety of pro-biotic cultures. The most common culture used for nut cheeses is rejuvelac, a fermented grain drink. Traditionally it has been made from wheat, but it can be made equally well with quinoa or rice as a gluten-free option. Rejuvelac can be found either with the

fermented beverages or the probiotics in your local natural food store. Store bought rejuvelac will be made from wheat. You can easily make your own using the recipe below.

If you don't want to use rejuvelac, there are many other cultures that will work. You can use commercial probiotics in liquid or powdered (pill) form. Choose one that has a broad spectrum of living bacteria. You may also want to test it before investing in it, as some probiotic blends give a very sulfuric taste and aroma (more gassy, than yummy). Water kefir, which is a SCOBY (symbiotic culture of bacteria and yeast) also works well for culturing nut cheese and gives it a tang. Test each of these out with small batches and see how you like the resulting flavor. For culturing non-dairy yogurts, I recommend using a plain soy based yogurt, pro-biotic powder or a commercially available vegan yogurt starter *(see Appendix A Resources)*.

Make Your Own Rejuvelac

Rejuvelac is a sour fermented grain beverage that contains living probiotic organisms. It takes several days to make at home, so for a fully DIY nut-cheese experience, you will want to start your rejuvelac a week or so before you are going to start your cheese. You will be capturing wild probiotic strains from your local environment. Once made, the rejuvelac will stay viable for 3-4 weeks in the refrigerator

The cloth cover, affixed with an elastic band allows the natural bacteria access while keeping out dust and flies.

1. Sprout the grains. Place one cup grains in a quart jar and add enough water to cover them. You can use wheat berries, rye berries, unprocessed brown rice, quinoa, whole oats, barley, or millet. Cover the mouth of the jar with a fine cloth and secure with a rubber band, or get a seed sprouting screen available at a natural food store. Let the

grains soak 8-12 hours, then drain. Place in a dark spot out of the sun and rinse once or twice a day for 1-3 days until the grains just barely start to sprout (you will see the little tail of the root sticking out).

2. Culture the grains. After rinsing one last time, add three cups spring water or other non-chlorinated water to the jar. Cover with a fresh cloth and affix with a rubber band (do not use the lid—the mixture must be open to the air). Place in a warm spot out of direct sunlight for another 1-3 days. The water will get cloudy and the mixture will start to bubble or smell sour. Then it is ready! Strain the liquid off the grains into a clean jar and use immediately or refrigerate.

Additives

Tapioca Flour/Tapioca Starch

Tapioca starch and tapioca flour are the same product, though fineness and texture may vary from brand to brand. Tapioca is derived from the manioc or cassava plant. It consists primarily of carbohydrates and is low in fat. It does offer folic acid, omega 3 and omega 6 fatty acids and a small amount of fiber (one gram per cup as compared to 10 grams in a cup of wheat flour).

Kappa Carrageenan

Kappa Carrageenan is a long chain carbohydrate that is extracted from red seaweed. It is widely used in the food industry, for gelling, thickening, and stabilizing. There have been some health concerns about it, that it may cause some inflammation in the gut, but studies have been inconclusive. Carrageenan is not a regular grocery store item, but can be found at several online outlets. While there are other thickeners and gelling agents that can be used, such as agar or xanthan gum, Carrageenan is the only one that will re-soften when heated, making it ideal for meltable vegan cheeses.

Nutritional Yeast

Nutritional yeast is a deactivated (not living) yeast sold in the form of flakes or powder. It is a source of protein and vitamins, especially the B-complex vitamins (except B12), and is a complete protein. It is low in fat and sodium and is free of sugar, dairy, and gluten. Nutritional yeast adds complexity and cheesiness to the nut cheeses. It can be found in the bulk aisle of most natural food stores.

Nut Milk

Almond milk is a favorite, but cashews, hazelnuts, macadamia nuts, pumpkin seed, or hemp seed also make delicious milk. Macadamia nuts, hemp seed, and cashews do not require pre-soaking. All nut milk can be made with or without the addition of dates and vanilla. This recipe makes about 2-1/2 cups and may be easily scaled up or down. Nut milk will keep 4-7 days stored in the refrigerator

2-1/2 c water
1 c whole raw soaked nuts or seeds (cashew, macadamia and hemp do not require soaking)
3 pitted dates (optional)
1/2 t vanilla extract (optional)

If using almonds, hazelnut, or similar, pre-soak 8-12 hours. If using pumpkin seed, pre-soak 3-4 hours. If using dates, pre-soak along with the nuts. Drain and rinse the pre-soaked nuts. Place the nuts into a high-speed blender with 1-1/2 c water. Add the optional dates and/or vanilla for a slightly sweeter milk. Blend until smooth. Add the remaining cup of water. Blend again. Strain the milk through a nut milk or muslin bag, squeezing to get all the liquid out. May be used at room temperature or chilled. For a creamier consistency only add 1/2 cup of water at the end (2 cups of water total).

Start soaking your almonds and dates just before bedtime to have fresh almond milk on your morning cereal.

Chai Milkshake

by Heather Haxo Philips

2 cups almond milk
1 peeled banana
1/4 c dates, pits removed
1/4 t cinnamon
1/4 t cardamom
1 t rosewater (optional)

Place all ingredients in blender and blend until smooth. Serve Chilled.

Frozen Fruit Shakes

1 c nut milk

1/2 c frozen fruit

1-2T sweetener

1. Make the nut milk with vanilla and dates. At the same time as you begin soaking the nuts, put the fruit into the freezer. For each portion of milk shake, freeze one peeled banana, or 1/2 cup diced fruit. Strawberries, peaches, raspberries, apricots, nectarine, and melons all work well.

2. Add the nut milk and the frozen fruit to the blender and blend on high for 1 minute or until fruit is macerated. For a sweeter product add 1 T maple syrup or honey.

Frozen Chocolate-Coconut Milkshake

This delightful shake will hearken back to childhood. It requires planning ahead, or simply freezing in batches during the summer so you have frozen ingredients on-hand. This recipe makes one jumbo shake or enough for two people to enjoy a smaller portion. The chocolate may be left out for a vanilla coconut version.

1 c nut milk made with vanilla and dates, frozen

1/2 c frozen coconut milk (1 small can full fat coconut milk or 1/2 cup homemade coconut cream and fresh coconut water)

1/2 c coconut water, nut milk or plain water

1T sweetened dark chocolate powder or 1 T unsweetened powder with 2 T maple sugar or 1 T honey

1/2 t vanilla extract

1 dollop coconut cream *(optional)*

1. Make the nut milk with vanilla and dates.

2. Pour 1 cup of milk into ice cube trays and freeze over night.

3. Drain the liquid portion off the canned coconut milk. and freeze the thick part in ice cube trays over night. Reserve the coconut liquid for your shake.

4. Place all ingredients into a high-speed blender and blend until it is a milkshake consistency. Pour into chilled glasses and top with a dollop of whipped cream.

Non-Dairy Mochaccino

1-1/2 c nut milk
1-2 shots espresso
1 T chocolate powder
1/2 t vanilla
2 T sweetener (optional)
6 ice cubes

1. Place all ingredients into the blender and blend on high until creamy.
2. Sweeten to taste.
3. Top with cashew or coconut cream.

Sweet Cashew Cream

This delicious cream can be served as an alternative to whipped cream on pies or over fresh fruit for a healthy but decadent dessert.

1 c cashews, soaked 2-4 hours
1/4-1/2 c water
3-6 pitted dates (depending how sweet you want it)
1 t vanilla extract

1. Blend all ingredients in a high-speed blender until smooth and creamy.
2. Chill to serve.

Coconut Cream (fresh coconut)

1 Thai coconut
1-3 T sweetener

1. Open one young coconut—also called Thai coconut. *(See adjacent sidebar)* Pour off the coconut water and scoop out the meat. Remove any of the brown hull that is attached to the meat.
2. Place the meat into a high-speed blender and blend until creamy. Delicious on fruit or as a dessert topping. For a sweet tooth add a small amount of maple syrup or honey and vanilla.

Thai coconuts, also called young coconuts,

can be found in Asian markets and at produce stands.

Tip!

How to Open A Young Coconut

Young Coconut, also called Thai coconut is a fresh and tasty alternative to canned coconut milk. The meat can be used in several recipes including cocnut cream and coconut yogurt. Although they appear somewhat daunting, they are actually very easy to open, as long as you know the trick!

Turn the coconut on its side. Using a large cleaver or vegetable knife, carve away the soft while outer skin covering the pointy end of the coconut to expose the brown crown of the nut.

Whack the outer edge of the exposed crown of the coconut with the back corner of your cleaver.

Twist and flip up the cleaver to crack off the top, being careful not to splll the young cocconut water inside.

Now you are ready to pour off the coconut water and scoop out the meat.

Coconut Whipped Cream (canned coconut milk)

2 cans full-fat store bought coconut milk, preferably organic, chilled

2 t vanilla extract

1/8-1/4 c sweetener of your choice (honey, maple, sugar etc)

1. Chill the coconut milk in the refrigerator overnight.
2. Open the cans from the bottom and drain off the non-solidified liquid part of the coconut milk. This can be reserved for use in milkshakes and smoothies.
3. Place all ingredients into a bowl and mix with a hand held mixer until light and fluffy and small peaks form.

After chilling the canned coconut milk overnight, open from the bottom, pour out the liquid portion and put the solid fatty potion into your bowl. Whip with your favorite sweetener for a decadent dessert topping.

Cashew Sour Cream

1/2 c cashews soaked 2-4 hours

1/2 c rejuvelac or water kefir

1 t lemon juice

1/4 t salt

1. Blend the nuts with the rejuvelac or kefir water in a high speed blender until smooth and creamy.
2. Scrape into a bowl or jar and let sit at room temperature covered for 24-48 hours. The sour cream will expand and fluff up as it ferments, so leave enough space in the fermenting vessel to allow for that. If using a jar, leave it slightly cracked open so carbon dioxide from fermentation can escape.
3. When it has reached the desired tartness mix in the salt and lemon juice and refrigerate. The resulting cream will be quite thick and great for soups or burritos. For a thinner consistency mix with a bit or water, soy milk, or nut milk. Makes 1.5 cups.

Creamy Ranch Dressing

1/2 c Cashew Sour Cream *(above)*
1/4 -1/2 cup plain soy milk or water
1 clove garlic, pressed
1 T fresh chives minced
1 T fresh dill, minced, or 1 t dried dill
Salt and pepper to taste

1. Put the sour cream into a jar or bowl.
2. Add the soy milk or water a little bit at a time, mixing thoroughly until you get to a thick and creamy yet pourable consistency.
3. Add the rest of the ingredients and mix.
4. Serve over salad or as a dipping sauce for French fries or fresh vegetable crudités.

Cheesy No-Cheese Sauce #1

1 c cashews soaked 4-6 hours
2 t powdered onion
2 T nutritional yeast
1 t smoked paprika or ground chipotle pepper

1. Place all ingredients into a high-speed blender and process until creamy.
2. Sauce may be heated for use in macaroni and cheese or nachos or to be served over vegetables or rice.

Cheesy No–Cheese Sauce #2

1 c cashews
1 medium tomato
1 small red bell pepper, roasted, skin, and seeds removed
1/4 c lemon juice
1/4 c nutritional yeast
salt

1. Place cashews, tomato, red pepper, nutritional yeast, and lemon juice into a food processor or high-speed blender and blend until creamy, adding water as needed to keep the mixture moving.
2. Add lemon juice and salt to taste. Maybe used as a dip or heated and blended with pasta for a macaroni and cheese effect.

Cheesy Kale or Zucchini Chips

1 head of kale or 3 medium zucchini
1 batch of Cheesy No-Cheese Sauce #2

1. Pour cheese sauce into a large bowl.
2. Slice zucchini into 1/4" slices or break kale into medium sized pieces.
3. Place into bowl and mix to coat thoroughly with the sauce.
4. Place in a dehydrator and dehydrate at 105-115° for 8 hours or until crispy.

NON-DAIRY YOGURT

Yogurt is a delicious, nutritious living food. Store bought non-dairy yogurt is almost always full of stabilizers and non-food additives. Making yogurt at home ensures a simpler whole food product. I tested just about every non-dairy yogurt option I could think of to bring you an array of delightfully flavored and textured possibilities.

The principle behind all yogurt is the same. You'll be introducing, heat-loving themophilic probiotic microorganisms to a food source and keeping them warm (100°-115°) while they do the work of congealing the fats and proteins in the food, creating a soft curd. The food source used will determine the texture and quality of the curd. The more fats and sugars available, the thicker it will be. For sources that create a thin or bodiless yogurt there is the option of adding a high-quality thickener. Depending on the base used for the yogurt, tapioca starch may work well, or you can use a high-quality vegan or bovine gelatin.

The thermophilic cultures come in several forms all of which work equally well. You can use a plain non-dairy yogurt from the store as your starter, a probiotic in powder or capsule form, or thermophilic culture. The thermophilic cultures used in the dairy cheese world work fine, but are raised on a milk substrate, so for those that wish to be 100% dairy free, vegan yogurt cultures are also available. They contain the same micro-organisms as the dairy culture, but are grown on a vegan base *(see Appendix A Resources)*.

Most plain non-dairy yogurt can be used as your starter culture, regardless of whether you are culturing soy, nuts, or coconut.

Soy Milk Yogurt

This recipe requires access to a good quality plain, store bought soy milk. Look for one that contains just soybeans and water. Both Eden Soy and

West Soy work well. The resulting yogurt is light and silky, but doesn't have a lot of body. For a thicker yogurt the addition of tapioca starch or gelatin works equally well.

1 quart plain soy milk
1/4 c plain soy yogurt or 1 capsule of probiotic powder or 1/8 tsp thermophilic culture or 1/8 t vegan yogurt starter
5 half pint jars
1/4 c tapioca starch, unflavored vegan jel or high-quality bovine gelatin *(use amounts listed on package)*

1. Heat the soy milk to 110°.
2. If you are using powdered culture, sprinkle it on top of the warmed milk. Let it sit for a half minute to dissolve, then stir in thoroughly. If you are using a prepared yogurt as your starter, mix it first with a 1/4 cup of the warmed milk, to break up the curds, then return the mixture to the pot and mix in thoroughly.
3. If you are using the optional tapioca starch or gelatin, sprinkle it over the milk, then mix in vigorously.
4. Pour the inoculated milk into the mason jars, distributing evenly.
5. Incubate at 100°-115° for 4-6 hours until congealed *(see Incubation Made Easy in the dairy yogurt section)*.
6. Place in the refrigerator over night to finish setting the yogurt. If you will be eating the yogurt within the week, you can add sweetener and vanilla directly to the milk before inoculating. Other flavors such as fruit should be added at the time of eating *(see Flavoring Yogurt, page 35)*.

Soy Milk Cashew Yogurt

The portions on this are loose. You can use any amount of cashews, up to 1/3 of the total volume to increase the creaminess and body of the yogurt.

3 c soy milk
1/4 c to 1-1/2 c cashews, soaked up to 8 hours

1. Place the cashews in the blender along with a small amount of the soy milk to start the blending process. Continue adding milk until the cashews are completely creamy.
2. Add the rest of the milk and blend another 5 seconds.
3. Heat the mixture to 105°-110°.
4. If you are using powdered culture, sprinkle it on top of the warmed milk. Let sit for a half minute to dissolve, then stir in thoroughly. If you are using a prepared yogurt as your starter, mix it first with a 1/4 cup of the warmed milk, to break up the curds, then return the mixture to the pot and mix in thoroughly.
5. Pour the inoculated milk into the mason jars, distributing evenly.

Incubate at 100°-115° for 4-6 hours until congealed *(see Incubation Made Easy, page 35).*

6. Place in the refrigerator over night to finish setting the yogurt. If you will be eating the yogurt within the week, you can add sweetener and vanilla directly to the milk. Other flavors such as fruit should be added at the time of eating (see *Flavoring Yogurt, page 35*).

Whole Nut Yogurt

1 c almonds or cashews, soaked 3-8 hours.

1/4 c plain non-dairy yogurt, 1 capsule pro-biotic powder or 1/16 t thermophilic culture

1. If using almonds, slip the skins off.
2. Place nuts into a high-speed blender adding just enough water, a little bit at a time, to keep the mixture moving. Continue to add water until the nut cream is the consistency of a thick lentil soup.
3. Blend in the culture as described in the previous two recipes.
4. Pour the mixture into sterile jars and incubate at 100°-115° for 4-8 hours, until thick.
5. Refrigerate to finish the gelling process.

Walnut Fig Yogurt

Absolutely delicious. And purple!

1 c walnuts, soaked 3-8 hours

3-5 medium figs

1/4 c plain soy yogurt, 1 capsule of probiotic powder, 1/16 tsp thermophilic culture or 1/8 t vegan starter

1. Blend nuts, figs and starter in a high-speed blender until creamy. Add water as you go to reach a thick, but pourable consistency.
2. Pour into sterile mason jars.
3. Incubate at 100°-115° for 4-8 hours until thick.
4. Refrigerate to finish the gelling process.

Figs and walnuts are the base for this wonderful purple yogurt, though blueberries, kiwis and pineapple guava work equally well.

Thai Coconut Yogurt

This yogurt requires a bit more effort and is more expensive than some of the others, but is well worth it as an occasional treat. The fresh coconut water alone can be cultured but requires the addition of gelatin to create a satisfying curd.

1 Thai coconut
1T non-dairy yogurt
Unflavored vegan jel or high-quality bovine gelatin *(use amounts listed on package)*

1. Open the young coconut and drain the milk *(See sidebar page 93)*
2. Scoop out the meat of the coconut and scrape off any of the brown skin that comes with it.
3. Put the coconut meat into the high-speed blender with a little of the milk and blend until creamy, adding coconut water as you go to get the consistency of a thick pourable paste.
4. Add the yogurt starter and optional gelatin and blend in well.
5. Place in sterile jars and incubate at 100°-115° for 4-8 hours until thick.
6. Place in the refrigerator to complete the gelling process.

Almond Milk Yogurt

This yogurt only works with the addition of high-quality gelatin, bovine or vegan. The yogurt will sometimes separate in the gelling process, much as homemade almond milk separates. If this is the case, simply mix it all back together when you serve it.

1 quart homemade almond milk with dates and vanilla *(see recipe page 90)*
1/4 c plain soy yogurt , 1 capsule of pro-biotic powder, 1/8 t thermophilic culture or 1/8 t vegan starter
Unflavored vegan jel or high-quality bovine gelatin *(use amounts listed on package)*

1. Heat the almond milk to 110°.
2. If you are using powdered culture, sprinkle it on top of the warmed milk. Let it sit for a half minute to dissolve, then stir in thoroughly. If you are using a prepared yogurt as your starter, mix it first with a 1/4 cup of the warmed milk, to break up the curds, then return the mixture to the pot and mix in thoroughly.
3. Sprinkle the gelatin over the milk, then mix in vigorously.
4. Pour the inoculated milk into the mason jars, distributing evenly.
5. Incubate at 100°-115° for 4-6 hours until congealed *(see Incubation Made Easy page 35)*.
6. Place in the refrigerator over night to finish setting the yogurt.

Yogurt Cream Cheese

Yogurt Cream Cheese, also called Yogurt Cheese is yogurt which has been strained through a cloth for a spreadable consistency. For best results use soy or soy-cashew yogurt without the addition of tapioca or gelatin. Be sure to refrigerate the yogurt for a few hours so that it is well set before continuing. Line a strainer with a nut milk bag or doubled fine cheesecloth. Pour in the yogurt and let it drain for 12-24 hours until it reaches a creamy, spreadable consistency. Flavor with a little salt and lemon for a product you can spread on your bagel. Flavor with lemon juice and a bit of your favorite sweetener for a cream cheese frosting.

Lebnah

1 quart soy milk cashew yogurt
1T lemon juice
1/2 t salt
1/4 c fresh dill, minced, or 1T dried dill
1/4 c olive oil

1. Make the yogurt with at least two parts soy milk and one part cashew. Refrigerate for a few hours before continuing.
2. Once the yogurt is set, drain for 24 hours in a strainer lined with a nut milk bag until it is a creamy paste.
3. Mix in lemon juice and salt. Sprinkle your working surface with dill and roll the cheese into a log, catching the dill on the surface.
4. Place on a plate, make a depression in the top of the cheese and fill with olive oil. Sprinkle the plate with any remaining dill.

Almond Ricotta

This creamy cheese can be flavored savory or sweet and can be used as a topping for dessert, the base on a cream pie or with salt garlic and herbs as a dip or spread.

1 c almonds, 1 c macadamia nuts or combination, soaked 3-8 hours
Water

1. If using almonds, slip the skins off.
2. Put the nuts in a high-speed blender with a small amount of water. Add a little water while blending as needed, but not so much the cheese becomes watery.
3. Scrape out, flavor, and serve *(see Flavoring Nut Cheese page 104)*

Pine Nut Parmesan – Dry Method

1 c pine nuts
1 c nutritional yeast
1/2 t sea salt

Grind until mixed and somewhat chunky. Do not over-process.

Tip: Grinding Nuts In A Highspeed Blender

For good texture it is important that the nut cheese be ground to a creamy consistency with as little liquid as possible to reduce the amount of draining needed. As the liquid is absorbed by the nuts, the blender blades push the nut mass out of the way and stops the blending process. This can be avoided by adding more liquid, and you will need to continue to add small amounts of liquid as the you go, but as already stated, unless you are making milk or cream you want to use as little liquid as possible.

To keep the nuts blending without adding too much liquid, use a spatula and carefully run it down the side wall of the blender, with the blades running. Push the spatula against the top of the nut mass with a little tap, this should push the nut mass towards the blades, without catching your spatula. Continue to push the nuts into the blades, tapping at the very top of the nut mass, keeping your spatula against the wall of the blender for the best control.

∽ CULTURED NUT CHEESES ∾

Fresh fermented nut cheeses are made primarily with almonds or cashews and favored or shaped in a variety of ways. Other nuts can be used or added into the mix for interest and flavor variation. Substitute 1/4 of your total with pumpkin seed for a greener, nuttier flavor or with macadamia or pine nuts for a richer, creamier mouth-feel. To culture you can use homemade or store-bought rejuvelac, an equal amount of water kefir or water plus 1/8 teaspoon probiotic powder (or two capsules emptied out). These tangy cheeses take a day or two to culture—anywhere from 12-48 hours depending on ambient room temperature and how tangy you like your cheese. Taste it as it cultures and stop culturing when you like the flavor. Refrigerating the cheese will slow down the culturing process, but not stop it all together. The cheeses will keep refrigerated for about a week. The longer they sit, the tangier they get. They may also be frozen for up to a month.

Almond Chevre

2 c almonds
1/2 c rejuvelac or water kefir or 1/2 c non-chlorinated water plus 2 pro-biotic capsules

1. Soak almonds briefly in near boiling water until the skins easily slide off. Slip the skins off all the almonds.
2. Put the almonds and the culture into a high-speed blender and blend until smooth and creamy. You may need to add a bit more liquid to get a really fine grind.
3. Scoop the pureed mixture into a nut cheese bag and hang to drain 12-24 hours.
4. Flavor your cheese using one of the options below.
5. Roll into logs with your hands or using wax paper and eat.
6. For a firmer cheese, flavor first then press. To press, line a small cheese press with a nut bag and press the cheese under light pressure for 3-6 hours. A jar of water or a stack of books work well as weights.

Creamy Cashew Cheese

2 c cashews, soaked 3-6 hours.

1/4 - 1/2 c rejuvelac or water kefir or an equal amount of non-chlorinated water plus 2 probiotic capsules emptied out

1. Place the cashews into the blender, plus just enough rejuvelac to process the cashews (the longer the cashews have soaked, the less liquid will be needed). Blend until smooth and creamy.

2. Transfer to a bowl or jar and let sit covered for 24-48 hours. If fermenting in a jar, do not cap too tightly, as you do not want the jar to explode.

All cultured nut cheeses will benefit from the addition of salt and lemon juice to bring out the flavor.

3. Once the cheese is tangy enough for your taste, flavor, and serve *(see flavoring suggestions below).*

Flavoring Nut Cheeses

All nut cheeses will benefit from the minimum addition of lemon juice and salt to taste. Add 1-2 t lemon juice and 1/4-1/2 t salt per cup of cheese. For a stronger cheese flavor, add 1T Nutritional yeast flakes, and/or 1T miso per cup of cheese.

To add your special flair, the following items can be mixed into your cheese before forming into a log or pressing into a bowl to shape it.

Herbs. Any combination of fresh or dried: basil, rosemary, thyme, marjoram, tarragon, or dill.

Alliums. Minced garlic, onion, shallots, or chives.

Pesto. Minced olives, or minced sun-dried tomatoes.

Lemon Zest.

Pepper.

Sweetener. Maple syrup, honey, coconut sugar or agave syrup.

Fruit. Diced or pureed.

Oil. For a creamier cheese mix in 1/2 cup olive or coconut oil. Press into a form, refrigerate. Flip cheese out of the form to serve.

Cheese Torte with Olive Tapenade, Sun-dried Tomato Pâté, or Pesto

For a beautiful presentation make a layered cheese.

1. Prepare or purchase one of the spreads below.
2. Line a small bowl or cheese press with a nut milk bag or cheesecloth.
3. Press in half the cheese. Spread a layer of the tapenade, pâté or pesto on top of that, and then a final layer of cheese.
3. Press firmly with a spoon, refrigerate 1-2 hours, then flip onto your serving plate.

Gather your materials, starting with one full batch of Almond Chevre *(see page 104)*

Next spread a layer of tapenade, pesto, or even fruit compote into the form.

Line the cheese form or bowl with cheesecloth or a piece of nut milk bag and spoon in half of the cheese.

Spoon in the other half and press with a spoon.

Olive Tapenade

1/2 pound pitted olives
2 cloves garlic
1 T olive oil
1 T lemon juice
a few leaves of basil or thyme

Blend 30 seconds in a food processor

Sun-Dried Tomato Pâté

1/2 c soaked sun dried tomatoes
1 medium red pepper, roasted.
1T olive oil
A few leaves of basil, thyme, or rosemary
salt and pepper to taste

Blend 30 seconds in a food processor

Pesto

2 bunches basil, de-stemmed
1-2 cloves garlic
1/2 c sunflower seeds, cashews or pine nuts
1/4-1/2 c olive oil
Salt to taste

1. Place basil, garlic, and seeds into a food processor.
2. Process adding oil slowly as needed to keep the mixture moving. The pesto should be creamy, but not runny.
3. Add salt to taste.

Boursin

2 c Almond or Cashew Cheese
1T lemon juice
2T nutritional yeast
2 cloves of garlic minced
1T each of fresh herbs: parsley, tarragon, marjoram, basil
A spring of fresh thyme or a pinch of dried thyme
1 t salt

1. Blend ingredients with a fork in a bowl.
2. Form into a mound and serve

Refrigerate for several hours then flip onto your plate

Sharp Cheddar

2 c almond or cashew cheese
1T lemon juice
2T Nutritional yeast
2-3 T red miso

1. Blend ingredients in a bowl or food processor.
2. Press into a square container and refrigerate.

Double-Cream

This makes a decadent creamy cheese that is a wonderful addition to a cheese platter.

1 c creamy cashew cheese
1/2 c coconut oil
2t lemon juice
salt
nutritional yeast
fresh herbs or freshly cracked pepper

1. Place everything but the herbs and pepper into a food processor and mix thoroughly.
2. Form the cheese into a ball. Sprinkle herbs and pepper onto a plate and roll the ball onto it to coat the outside.
3. Press into a cheesecloth or nut milk bag lined bowl and refrigerate.
4. Plate the cheese and let it sit at room temperature one hour before serving.

Pine Nut Parmesan—Cultured Dehydrator Method

by Heather Haxo Philips

3/4 c water
1-1/2 c raw pine nuts
3/4 t salt
2 probiotic capsules emptied out

1. Add all ingredients to high-speed blender and blend until smooth.
2. Spread the mixtures onto teflex lined dehydrator trays and dehydrate at 105°-110° degrees until dry and flaky.
3. Once completely dry, skim or peel the cheese off the teflex and break up into flaky crumbles.
4. Use immediately or store tightly covered in the refrigerator.

MELTABLE VEGAN CHEESE

Unlike the cultured nut cheeses we have learned about so far, meltable cheeses require cooking and are therefore not a living food. They also require a thickener that will allow the mixture to re-soften when heated to get the "melty" effect. The thickener that works in this way is Kappa Carrageenan, an additive which is "natural" yet highly processed *(see additives, page 89, for more detail)*. Meltable cheeses will also require an additional thickening agent such as tapioca flour. While again a "natural' product, tapioca flour is a highly processed carbohydrate. If you are non-dairy primarily for health and a whole foods diet, you may not want to enter the world of meltable non-dairy cheese. On the other hand, many of us crave that melty sensation, if not for the body, then for the spirit. Much of the research done on good quality meltable vegan cheese was done by author Miyoko Schinner. These recipes are adapted from her research. If you want more in this direction I highly recommend her book, *Artisan Vegan Cheeses.*

Quick Melty Mozzarella

This is a softer cheese which will slice and melt well in grilled sandwich or on top of a casserole.

1 c non-dairy yogurt
1/2 c water
1/4 c tapioca starch
1/3 c cold pressed olive or canola oil
1 T nutritional yeast
2T lemon juice
1/2 t salt
1T kappa carrageenan

1. Place all ingredients into a high-speed blender. Blend until well mixed and creamy.
2. Transfer the mixture into a small saucepan. Cook on medium heat stirring constantly with a whisk until the mixture thickens and pulls away from the sides of the pot.
3. Pour the thickened mixture into a glass bowl and allow to cool at room temperature.
4. Cover and place in the refrigerator until firm, 2-3 hours.

Quick Melty Mild Cheddar Cheese

This is a softer cheese which will slice and melt well in grilled sandwich or on top of a casserole.

1 c non-dairy yogurt
1/2 c water
1/4 c tapioca starch
1/3 c cold pressed olive or canola oil
2 T nutritional yeast
1-2 T red miso depending how strong a flavor you like
1/2 roasted red pepper, no seeds
1/2 t salt
1T kappa carrageenan

1. Place all ingredients into a high-speed blender. Blend until well mixed and creamy.
2. Transfer the mixture into a small saucepan. Cook on medium heat stirring constantly with a whisk until the mixture thickens and pulls away from the sides of the pot.
3. Pour the thickened mixture into a glass bowl and allow to cool at room temperature.
4. Cover and place in the refrigerator until firm, 2-3 hours.

Oat American Cheese

By Miyoko Schinner

2-1/2 c water
1 c rolled oats
2/3 c nutritional yeast
1 t lemon juice
3T miso
1/2 roasted red bell pepper, skinned
1/2 t salt
1 t mustard
1/2 c coconut oil or olive oil
1T agar powder
1T kappa carrageenan

1. Put the water and oats into a saucepan to cook the oats. Once the water comes to a boil, simmer the oats stirring frequently until thick, about 5 minutes.
2. Place the cooked oats in a high-speed blender along with all other ingredients and blend until smooth and creamy.
3. Transfer to a saucepan and cook over medium heat, stirring constantly with a whisk, 4 or 5 minutes until thick.
4. Pour the thickened mixture into a glass bowl and allow to cool at room temperature.
5. Cover and place in the refrigerator until firm, 6 hours.

Sometimes you just need a melty gooey cheesy experience

The quick melty cheddar also slices up quite nicely.

App

endices

APPENDIX A

Resources

Books

Artisan Cheesemaking at Home by Mary Karlin

If you only want to buy one book, I recommend this one. She approaches the book like a cheese course, starting with easier projects and building skills to more difficult cheeses.

200 Easy Homemade Cheese Recipes by Debra Amrein-Boyes

This book has great recipes including many artisan cheeses. The recipes are thorough and easy to follow. This is the only book I have found that has especially good troubleshooting tips. A downside is that later recipes in the book require 16 -20 quarts of milk. Cheese recipes are sometimes scalable and sometimes not, so you would have to take that risk.

Home Cheesemaking by Ricki Carroll

The recipes in this book are excellent, but many concepts are not fully explained and some recipes call for direct-set cultures which will confuse the beginner by making you think you have to buy a special culture (most of them are pre-measured mesophilic culture). Ricki Carroll held the torch of home cheesemaking for many years and runs New England Cheesemaking Supply.

Cheesemaking Made Easy by Ricki Carroll

Ricki Carroll's earlier book is out-of-print but can be found online. It is charming and more user friendly for the beginner, as she was herself, closer to being a beginner.

Goats Produce Too! The Udder Real Thing Vol. II By Mary Jane Toth

This lovely gem of a book focuses on goat cheeses but the recipes also work for cow. Ms Toth is a no-nonsense farmsteader who lists just a few tried and true recipes, as well as many ideas of how to use the simple cheeses you produce. Highly recommended.

The Untold Story of Milk
by Ron Schmid, ND
A wealth of information about the history, science, health and politics of milk. Well documented research and information on raw milk.

Artisan Vegan Cheese
by Miyoko Schinner
The result of deep research into vegan cheesemaking this book is the current bible on the topic. Many of the recipes call for thickeners and additives or for cooking the nuts, but the work is impressive it its scope.

The Art of Fermentation
by Sandor Katz
Everything you ever wanted to know about fermentation included many recipes for fermented dairy and nuts.

Websites

Fankhauser's Cheese Course
Google "Fankhausers Cheese Page Beginning cheesemaking"

Fiasco Farms
www.fiascofarm.com books, recipes, cultures

Cheese Forum cheeseforum.org/articles

Artisan Vegan Life
http://www.artisanveganlife.com
Miyoko Shinners website and blog

Supplies:

New England Cheese Supply
www.cheesemaking.com

The Cheesemaker
www.thecheesemaker.com

The Dairy Connection
www.dairyconnection.com

Glengarry Cheesemaking
www.glengarrycheesemaking.on.ca

Hœgger Supply
www.hoeggerfarmyard.com

Yemoos
www.yemoos.com
Kefir culture, Viili, and others

GEM Cultures
www.gemcultures.com
Kefir, Viili, Fil Mjolk, and others

Cultures for Health
www.culturesforheaslth.com
A variety of yogurt cultures including vegan

Living Nutz
www.livingnutz.com
raw organic nuts

Rabbit Food Grocery
www.rabbitfoodgrocery.com
Unflavored Vegan Jel

APPENDIX B

Which Critters Are in the Culture?

Mesophilic Culture

Lactococcus lactis ssp lactis and Lactococcus lactis ssp. Cremoris are the two main mesophilic strains which produce lactic acid at temperatures ranging from 77° - 90° They are used alone or in a blend for most cheese ripened at the above temperatures such as cheddar, Colby, Chevre, Feta, Brie, etc. Lactococcus Lactis ssp. Diacetylactis ferments at a faster rate producing large amount of carbon dioxide gas and diacetyl. It is used in mesophilic blends for cheese where an open structure is desired such as blue cheese or havarti.

Mesophilic Starter Series MM100-101

(LL) Lactococcus lactis subsp. lactis
(LLC) Lactococcus lactis subsp. cremoris
(LLD) Lactococcus lactis subsp. lactis biovar diacetylactis

This culture can also be used to make Camembert/brie, blue cheese, mozzarella, ricotta, chevre, and others.

Mesophilic Starter Series MA11

(LL) Lactococcus lactis subsp. Lactis
(LLC) Lactococcus lactis subsp. Cremoris

This culture will add a milder flavor to your Camembert/brie). Can also be used to make cheddar, colby, Monterey Jack, Blue cheese, Feta, Chevre, and others.

Mesophilic Starter Series MA 4000-4001

(LL) Lactococcus lactis subsp. Lactis
(LLC) Lactococcus lactis subsp. Cremoris
(LLD) Lactococcus lactis subsp. lactis biovar diacetylactis
(ST) Streptococcus thermophilus

Used for a variety of hard and semi-hard cheeses including Roquefort, Cambozola, Castle Blue, Colby, Cheddar, Gouda, and Brick cheeses with all types of milk. Also excellent for Camembert & Brie. The MA 4000 series blends both standard mesophilic lactic acid cultures plus a thermophilis for quicker acid production during cheesemaking.

Thermophilic Cultures

Streptococcus thermophilus, Lactobacillus delbruecki ssp bulgaricus and Lactobacillus delbruecki ssp helveticus are the main strains of thermophilic bacteria which produce lactic acid at temperatures ranging from 95° to 105° and can survive up to 140°. They are used in yogurt making and hard southern cheeses such as Parmesan, Romano, Emmental, and Gruyere.

Thermophilic Starter Series TA61

(ST) Streptococcus thermophilus

For making hard, Italian & Swiss cheeses like Parmesan, Romano, Provolone, Mozzarella, and Emmental/ Swiss.

Thermophilic Cultures Type B

(ST) Streptococcus thermophilus

(LBB) Lactobacillus delbruecki subsp. bulgaricus

For the production of Italian style cheeses, such as Mozzarella, Romano, Parmesan, Provolone, and Grana (Reggiano).

Thermophilic Cultures Type C

(ST) Streptococcus thermophilus

(LBB) Lactobacillus helveticus

For the production of all Swiss style cheeses, such as Gruyere, Swiss, Jarlsberg, and Emmental.

Direct Set Cultures

Direct Set Cultures are pre-measured doses of culture with rennet added.

Listed below are the ingredients of some of the more common direct set blends:

Buttermilk s.lactis, s.cremoris, l.b.diaetylactis, m.s.cremoris, malto dextrin.

Chevre s.lactis, s. cremoris, s. lactis biovar diacetylactis, malto dextrin, and vegetable rennet.

Crème Fraîche s.lactis, s. cremoris, s lactis biovar diacetylactis, malto dextrin and vegetable rennet.

Fromage Blanc, s.lactis, s.creamoris, l. biovar diacetylactis, vegetable rennet, and malto dextrin

Bulgarian Yogurt (ST) s.thermophilus, (LB) l.bulgaricus, skim milk, lactic cultures, ascorbic acid.

APPENDIX C

All About Rennet

Rennets vary in strength and their ability to coagulate also varies with the type of milk being used. Keep liquid rennet in refrigerator. Keep powder and tablet rennet in the freezer.

Rennet. A natural complex of enzymes produced in any mammalian stomach to digest the mother's milk, including a proteolytic enzyme (protease) that coagulates the milk, causing it to separate into solids (curds) and liquid (whey). The active enzyme in rennet is called chymosin or rennin but there are also other important enzymes in it, e.g., pepsin, or lipase. There are non-animal sources for rennet that are suitable for vegetarian consumption.

Natural calf rennet. Natural calf rennet is extracted from the inner mucosa of the fourth stomach chamber of young, unweaned calves. Thus it is a by-product of veal production. If rennet is extracted from older calves it contains less or no chymosin, but a high level of pepsin and can only be used for special types of milk and cheeses. As each ruminant produces a special kind of rennet to digest the milk of its own mother, there are milk-specific rennets available, such as kid goat rennet especially for goat milk and lamb rennet for sheep milk.

Traditional method. Dried and cleaned stomachs of young calves are sliced into small pieces and then put into saltwater or whey, together with some vinegar or wine to lower the pH of the solution. After some time the solution is filtered. The crude rennet that remains in the filtered solution can then be used to coagulate milk. Today this method is used only by traditional cheese-makers in central Europe.

Modern method. Deep-frozen stomachs are milled and put into an enzyme-extracting solution. The crude rennet extract is then activated by adding acid. After neutralization of the acid, the rennet extract is filtered in several stages and concentrated until reaching a typical potency of about 1:15000; meaning 1 gram of extract would have the ability to coagulate 15000 grams (15 liters) of milk. Typically, one pound of cheese contains about 0.00015 grams of rennet enzymes.

Vegetable rennet. Many plants have coagulating properties. Homer suggests in the Iliad that the Greeks used an extract of fig juice to coagulate milk. Other examples include nettles, thistles, mallow, and Ground Ivy. Enzymes from thistle are used in some traditional cheese production in the Mediterranean. Worldwide, there is no industrial production of vegetable rennet. Commercial so-called vegetable rennets are usually microbial rennet.

Microbial rennet. Some molds such as Rhizomucor miehei are able to produce proteolytic enzymes. These molds are produced in a fermenter and then specially concentrated and purified to avoid contamination with unpleasant byproducts of the mold growth. At the present state of scientific research, governmental food safety organizations such as the EFSA deny safety status to enzymes produced by these molds. The flavor and taste of cheeses produced with microbial rennets tend towards some bitterness, especially after longer maturation periods. Organic Vegetarian Rennet is a form of microbial rennet that adheres to organic standards.

Genetically engineered rennet. Because of the imperfections and scarcity of microbial and animal rennets, producers sought replacements. With the development of genetic engineering, it became possible to extract rennet-producing genes from animal stomachs and insert them into bacteria or yeasts to make them produce chymosin during fermentation. The genetically-modified microorganism is killed after the chymosin is isolated from the fermentation broth, so that the Fermentation-Produced Chymosin (FPC) used by cheese producers does not contain any GM ingredient and thus it is not listed on any label as such.

Chymosin produced by genetically modified organisms was the first artificially produced enzyme to be registered and allowed by the U.S. Food and Drug Administration. In 1999, about 60% of U.S. hard cheese was made with genetically engineered chymosin and it has up to 80% of the global market share for rennet. By 2008, approximately 80-90% of commercially made cheeses in the United States were made utilizing GMO-based rennet, whereas cheese from Europe is more likely to be made from animal rennet due to tradition. GMO rennet is technically vegetarian, as it contains no animal product, however ultimately it is still sourced from animals.

SALT

Homemade Cheese Record Form

Type of Cheese ____________________

Date ____________________

Type of Milk ____________________

Amount of Milk ____________________

1. Ripening

Type of starter ____________________

Amount of starter ____________________

Time at adding starter ____________________

Milk temperature at time of adding starter ____________________

2. Renneting

Type of Rennet ____________________

Amount of Rennet ____________________

Time at adding rennet ____________________

Milk temperature at time of adding rennet ____________________

3. Cutting The Curds

Size of Curds ____________________

Time at cutting the curds ____________________

4. Cooking The Curds

Time at start of cooking ____________________

Temperature at start of ____________________

Temperature at finish of cooking ____________________

Length of time cooking ____________________

5. Draining The Curd

Time at start of draining ____________________

Length of time draining ____________________

6. Milling the curd

Time of milling ____________________

7. Salting the curd

Amount of salt added ____________________

Type of salt added ____________________

Type of herbs added ____________________

Amount of herbs added ____________________

8. Pressing the curd

Time at start of pressing ____________________

Amount of pressure at start ____________________

Amount of pressure at end ____________________

Total time pressing ____________________

9. Air drying / Mold Ripening

Date started ____________________

Date ended ____________________

10. Waxing

Date waxed ____________________

11. Brining

Amount of salt added per pint ____________________

12. Washing

Type of solution used ____________________

How often washed ____________________

13. Aging

Temperature of aging ____________________

Start date of aging ____________________

End date of aging ____________________

14. Eating

Date of first bite ____________________

Taste of first bite ____________________

NOTES

Homemade Cheese Record Form

Type of Cheese ______________________

Date ______________________

Type of Milk ______________________

Amount of Milk ______________________

1. Ripening

Type of starter ______________________

Amount of starter ______________________

Time at adding starter ______________________

Milk temperature at time of adding starter ______________________

2. Renneting

Type of Rennet ______________________

Amount of Rennet ______________________

Time at adding rennet ______________________

Milk temperature at time of adding rennet ______________________

3. Cutting The Curds

Size of Curds ______________________

Time at cutting the curds ______________________

4. Cooking The Curds

Time at start of cooking ______________________

Temperature at start of ______________________

Temperature at finish of cooking ______________________

Length of time cooking ______________________

5. Draining The Curd

Time at start of draining ______________________

Length of time draining ______________________

6. Milling the curd

Time of milling ______________________

7. Salting the curd

Amount of salt added ______________________

Type of salt added ______________________

Type of herbs added ______________________

Amount of herbs added ______________________

8. Pressing the curd

Time at start of pressing ______________________

Amount of pressure at start ______________________

Amount of pressure at end ______________________

Total time pressing ______________________

9. Air drying / Mold Ripening

Date started ______________________

Date ended ______________________

10. Waxing

Date waxed ______________________

11. Brining

Amount of salt added per pint ______________________

12. Washing

Type of solution used ______________________

How often washed ______________________

13. Aging

Temperature of aging ______________________

Start date of aging ______________________

End date of aging ______________________

14. Eating

Date of first bite ______________________

Taste of first bite ______________________

NOTES

ENDNOTE

What Now?

If you have mastered the recipes in this book, Congratulations! You are well on your way to becoming a proficient home cheese maker. You now have all the basic skills and equipment to succeed with more ambitious cheeses. Your next step may be to seek out one of the books on the following resource page and try your hand at Jack, Cheddar or an aged, air-dried nut cheese. Or perhaps you are plenty satisfied with what you have learned here and will continue to hone and develop them for your own kitchen. Either way go forth proudly and impress your friends! Happy cheese making!

// ACKNOWLEDGEMENTS //

My thanks and appreciation go to:

Joe Biel and the crew at **Microcosm** for saying yes and for all their hours of work and dedication to making this and so many other awesome books.

Heather Haxo Phillips and **rawbayarea.org** for the inspiration for the title of this book and for teaching me to make my first nut milks and cheeses.

Miyoko Shinner for her valuable research on meltable vegan cheeses and for permission to use her recipe.

Jim Montgomery, my goat mentor and friend, for the invaluable experience of apprenticing with the goats at Green Faerie Farm, for suggestions to improve the ethical cheese section and for helping to stir the curds.

My urban dairy farming friends, **Jeannie McKenzie & Maya Blow**, for sharing in the home cheesemaking journey with all of its successes and disasters.

My partner, **Erik "Jesus" Bjorkquist** for seeming to enjoy everything I put on his plate and for saying yes to every wacky manufacturing project I propose.

This book is dedicated, in wonder, to the ***transformative action of micro-organisms***, and the mysterious ways they partner with us to make our human culture.

∽ ABOUT THE AUTHOR ∾

K. Ruby Blume is an educator, gardener, beekeeper, artist, performer and activist. A life-long learner, she has studied everything from permaculture to pollination ecology and has taught herself cooking, canning and fermentation techniques, as well as how to set tile, install a sink, do electrical wiring, tend a beehive and repair a motorcycle. Ruby holds certificates in permaculture design, holistic bodywork, sex education and coaching and has studied botany, herbal medicine, nutrition, microbiology and native plant ecology. She has extensive experience in the arts including work with ceramic, mosaic, glass, textile, printmaking, puppetry, collage, costume design and photography. She is known for her work as founder and artistic director of the art for justice project, Wise Fool Puppet Intervention, and has performed and exhibited her work since the 1980s. The product of three generations of teachers, Ruby's experience as an educator extends back thirty years. She has taught puppetry, theatre, art, gardening, stiltwalking, beekeeping, canning, cheese making and more to people ages five to ninety-five. She founded The Institute of Urban Homesteading in 2008 and co-authored the book *Urban Homesteading: Heirloom Skills for Sustainable Living*. (Skyhorse, 2011). Ruby maintains an urban farm on a 10th of an acre in Oakland, CA and is a proud generalist.

iuhoakland.com
rogueruby.com
ruby@sparkybeegirl.com

SUBSC... EVERYTHING WE PUBLISH!

Do you love

Do you wa
stuff?

Would you
as it's pub

Subscribe
we'll mai
released!

$10-30/m
Include
date of b
Subscri
purchas

...AN